# THE LAND
# OF FRANCE
# 1815~1914

The London Research Series in Geography

# THE LAND
# OF FRANCE
# 1815~1914

## Hugh D. Clout

*Department of Geography, University College London*

London
GEORGE ALLEN & UNWIN
Boston          Sydney

**George Allen & Unwin (Publishers) Ltd,**
**40 Museum Street, London WC1A 1LU, UK**

George Allen & Unwin (Publishers) Ltd,
Park Lane, Hemel Hempstead, Herts HP2 4TE, UK

Allen & Unwin Inc.,
9 Winchester Terrace, Winchester, Mass 01890, USA

George Allen & Unwin Australia Pty Ltd,
8 Napier Street, North Sydney, NSW 2060, Australia


First published in 1983

**British Library Cataloguing in Publication Data**

Clout, Hugh
   The land of France 1815–1914. – (The London
research series in geography; 1)
1. Land use, Rural – France – History
I. Title      II. Series
333.76'0944      HD645
ISBN 0–04–911003–9


**Library of Congress Cataloging in Publication Data**

Clout, Hugh D.
   The land of France, 1815–1914.
(London research series in geography, ISSN 0261–0485 ; 1)
Bibliography: p.
Includes index.
1. Land use, Rural – France – History.   2. Agriculture
– France – History.   3. Cadasters – France – History.
I. Title.   II. Series.
HD645.C56   1983      333.76'13'0944      82–13880
ISBN 0–04–911003–9


Set in 10 on 12 point Bembo by Nene Phototypesetters Ltd
and printed in Great Britain
by Mackays of Chatham

# Preface

My interest in the French countryside goes back to vacations as an under-graduate when I tramped through the lanes and fields of Normandy and Poitou with a notebook in my hand, a camera slung round my neck, and a tent stowed in the rucksack on my back. As well as recording the intricacy of fields, farms, barns and land use in the 1960s, I discovered the maps and registers of the *ancien cadastre* and came to realise that it would be possible to reconstruct aspects of the rural scene that had existed more than 100 years previously. A research studentship enabled me to investigate a much wider range of source materials in a study of the Pays de Bray from 1750 to 1965 and in subsequent years I turned part of my attention to investigating various aspects of change in the French countryside, sometimes working at the national scale, at other times framing my enquiries in particular regional contexts. Visits to archives in many parts of France demonstrated the rich potential (as well as the shortcomings) of 19th-century cadastral data, agri-cultural enquiries and professional writings for the rural geographer and encouraged me to make use of some of them in a cross-sectional study of *Agriculture in France on the eve of the railway age* (1980).

In that book I demonstrated the role of quantitative cartography in adding an areal dimension to complement all too familiar national averages and in providing a spatially sensitive statistical framework to which the results of regional and local monographs might be related. In short, I tried to provide a single datum plane of information which might be of interest to process-orientated researchers. In subsequent months I began to realise how ignorant I was of the varied pace of social, economic and landscape change in France's component provinces and *pays* during the 19th century. The point was hammered home by a couple of research trips to Gascony and to Provence which confirmed my growing feeling that the character and chronology of rural change in the Midi since the *ancien régime* had been substantially different from what I was familiar with in northern, western and central France. One particular visit to the upper valley of the Durance showed me agricultural landscapes that had been abandoned 120 years ago, while in Finistère land was being brought freshly into cultivation as recently as the 1920s!

*The land of France* explores the content of successive agricultural enquiries, cadastral revisions and a large number of official and academic reports in an attempt to elucidate my problem by reconstructing conditions at a series of dates between 1815 and 1914 and by demonstrating or inferring land-use changes that occurred during clearly defined intervening periods. In so doing I am well aware that the study of rural France in the 19th century may be likened to a well ploughed field, which already has yielded many rich harvests. In recent years a number of historians have probed the persistence of traditional *mentalités* and have exemplified the gradual and uneven transition from localism to national unification that occurred during the 19th century. Of late some historical geographers have alluded to the importance of politics and social conflict in comprehending the reorganisation of economic space that took place during that period but, although their argument is persuasive, their empirical contribution strikes me as being somewhat unconvincing. In my opinion, splendid theory has not yet been wedded effectively to hard evidence in this 'new' brand of historical geography. Undoubtedly such achievements will be made in due course. The present volume adopts a much more conservative stance in full recognition of the limitations of available sources. Nonetheless, it is written in the fervent hope that future scholars may find ways of bridging the tantalising gulf between general theory and partial empirical records.

Unlike my earlier book, which involved the reconstruction of a single chronological cross section, *The land of France* seeks to pursue a more explicitly dynamic, yet still modest, approach. The present work is focused quite deliberately on the study of land use; this choice having been made, first, in response to the uneven record of many other rural themes between 1815 and 1914, and secondly, with due deference to the opinions of French reviewers regarding the hazards of using 19th-century sources to calculate indices of food consumption by country dwellers or the financial yield of farming activities. Hence these themes and a number of other issues, such as commodity prices and crop combinations, are either omitted or treated with brevity in the present discussion. In addition, the scale of analysis is shifted from the *arrondissement*, which I employed in my earlier investigation, to that of the *département*. The reason for this is simple and inescapable: the département is the only unit below the nation as a whole for which data are available between 1815 and 1914. Unlike the presentation of *Agriculture in France*, reliance has been placed on the desk calculator and on pen and ink rather than the computer. This change is not because of any distaste for automated processing (quite the reverse) but just reflects the way that I have worked of late, with 'the book' being squeezed into spare half days, wet Sunday afternoons and the surprisingly productive amount of time spent on London's trains and buses.

I must record my great good fortune in having had the assistance of Christine Daniels who converted my cartographic drafts into publishable artwork; her skill and patience are truly amazing. I also express my gratitude to Chris Cromarty, Annabel Swindells, Claudette John and Debbie Ryan for their respective photographic and secretarial talents. Librarians and archivists in Paris, London and a number of French provincial cities helped me in many ways, although it has to be admitted that most of the detailed sources I consulted over the years fall beyond the scope of the present national overview. I have learned not to be surprised by the quantity of relevant material held by the Bibliothèque Nationale and the British Library, but the unclassified French holdings in the Whitehall library of the Ministry of Agriculture and the statistics conserved by the Direction Générale des Impôts (sous-direction du cadastre) in Paris were totally unexpected.

Over the years I have benefited from financial support for work in France from the Centre National de la Recherche Scientifique, the Social Science Research Council, the Central Research Fund (University of London) and University College London. Research for the present book and for a number of earlier publications has been made possible in this way. I extend my gratitude to my former Head of Department, Professor Emeritus Bill Mead, who provides a powerful example of what it means for a geographer to 'adopt other lands'. My colleagues and students (especially the faithful band taking the historical geography of France) continued to put up with me, providing just the right blend of criticism and stimulus in a congenial atmosphere. Last, but by no means least, my friends on both sides of the Channel provided welcome company and conversation in both town and country and managed to drag me away from archives, libraries and museums with collections of farm implements into restaurants, bars and cinemas. They were right, of course.

HUGH CLOUT
Bloomsbury, 1982

# Contents

# List of tables

# A note on place names

Names of *départements* are given in their 19th-century form (e.g. Charente-Inférieure, rather than Charente-Maritime) but where English versions exist for cities (e.g. Lyons), *pays* (e.g. Upper Normandy), provinces (e.g. Brittany), regions (e.g. Paris Basin) and other geographic phenomena, they have been preferred. Unlike the French convention, the preceding definite article has normally been omitted before the names of départements and pays. The unusual term 'middle France' is employed to describe a group of pays on the southern fringe of the Paris Basin, including Nivernais, Berry, Sologne, Touraine, Val de Loire and Brenne.

# 1   *Purpose and plan*

## Meshing space and time

At the dawn of the 20th century over half the people of France still lived in
the countryside and two-fifths of her workforce were engaged in farming
and forestry. All this was strikingly different from the situation in Great
Britain or Germany where towns and factories had claimed substantially
greater proportions of the national population. Each country, of course, had
pursued its unique path of economic and social development toward the
new century but France appeared to have lagged significantly behind her
neighbours (O'Brien & Keyder 1978). Generations of scholars have turned
their attention to this question, plotting trends and advancing new inter-
pretations to challenge conventional wisdom regarding the occurrence of
agricultural, industrial and urban 'revolutions' in 19th-century Europe and
in France in particular. Their arguments vary in emphasis but each tends to
point back to the nature of peasant landownership and farming activity as the
mechanism for retaining so many Frenchmen on the land for so long. When
set against the British or German models, change in the French economy was
not great, rapid or 'revolutionary' but was taking place all the same (Caron
1979).

During the course of the peaceful century between the Napoleonic Wars
and World War I many aspects of continuity from a more remote past
survived in the villages and farms of France. Old traditions lived on and
primary education made tardy progress in the depths of the countryside as
Eugen Weber has shown so convincingly in his masterly assemblage of
information entitled *Peasants into Frenchmen* (1977). Nonetheless, rural
France did experience substantial and sometimes devastating change during
those 100 years. Population numbers rose and fell, craft industries flourished
and foundered, migrants left the countryside for the town, and all aspects of
farming were exposed to outside commercial forces to a greater or lesser
degree as modern systems of transport were insinuated gradually and un-
evenly through the countryside. Individual *pays* and branches of production
were subjected to change at varying paces and in differing ways. Some were
transformed markedly and suddenly, others only slightly and slowly, but
none could remain completely unaltered. In the spirit of Meinig's (1978–9)
vision of America, 19th-century France may be thought of as 'an ever-
changing place, an ever-changing congeries of places, an ever-changing
structure of places, and an ever-changing system of places' (p. 1202). Not
surprisingly, the general pace of change was greater during the railway age
than ever before.

Quantitative historians have devised ingenious indicators to depict temporal fluctuations in the whole national economy and its component sectors, while other historical scholars have worked at a finer scale to trace the evolution of complete regional economies, of key aspects of manufacturing or, more rarely, of farming activity (Toutain 1961, Chaunu 1972). Following the tradition of Vidal de la Blache (1903), many French geographers have been attracted to studying rural life since interrelationships between human and physical environments seemed to be so much more explicit in rural areas than in towns. Despite the functionalist underpinnings of Vidalien thought, generations of geographical researchers provided ample testimony of both continuity and change in the countryside during the 19th century, and the closer the scale of analysis the more striking the changes that emerged. Like many of their historian colleagues, French geographers normally studied individual provinces or pays, rarely lifting their eyes to glimpse the wider horizons of France as a whole (Demangeon 1946). As a result of their labours the amount of knowledge about the structure and evolution of farming in many areas of France in the 19th century is truly vast, but works of synthesis which attempt to establish a spatially sensitive framework to embrace the whole nation are few (Klatzmann 1955). The areal approach that historical geographers such as myself might wish to advance and exemplify through cartography has rarely been extended to studying rural life throughout France (Braudel 1951).

Given this fact, it is not surprising that even fewer attempts have been made to employ such an areal approach in a dynamic way in order to portray and elucidate rural change across space and through time (Plaisse 1963). Neither national time-series nor detailed regional studies are appropriate media for comprehending the complex and intricate juxtaposition of climate and culture, pays and provinces that is France. For a foreigner, lacking the proud attachment that many French scholars possess to their home region, the absence of such a dynamic spatial approach is quite bewildering. The reasons for this state of affairs must be sought deep in the structure of French historical and geographical scholarship, where the regional monograph reigned supreme for many decades. More recently an apparent decline of interest in the past among French human geographers has also played its part. Work on regional disparities in agricultural growth by the economist Jean Pautard (1965) provides the only sustained attempt to mesh space and time in the way that some historical geographers might wish to do.

It cannot be argued that there is a sufficiently serious shortage of evidence to prevent undertaking such an enterprise since numerous attempts at national stocktaking between 1814 and 1914 offer an abundance of quantitative data, which forms rich and challenging quarry for the researcher, be he geographer or member of another discipline. These sources are, of course, far from faultless and the scope of issues they embrace is selective (Gille 1964). Their compilation ranges from simple estimation to rigorous local enquiry, their periodicity varies, and their systematic coverage is uneven. Normally they are records of static conditions at specific points in time rather than explicit expositions of processes at work, and hence the researcher is required to tread the hazardous path of inference as he seeks to determine directions and trends of change by juxtaposing two or more cross sections in time. But, of course, unravelling the past is never easy, even with the help of the vast cadastral surveys and agricultural enquiries which offer great potential for reconstructing the look of the land and tracing aspects of agricultural activity during the crucial transition phase that spans the final decades of the *ancien régime économique* and the early decades of the modern age.

The choice of the end of the Napoleonic Wars and the eve of World War I as appropriate bounding dates for studying this critical period may be justified in at least three ways. First, the intervening years formed a long and almost uninterrupted period of domestic peace. By 1815 a quarter century of disruption was drawing to a halt, while in 1913 Western Europe was on the brink of a more devastating war than had been known before. Echoing the words of Sir John Clapham (1921), these dates were 'the starting and finishing points for a great age' which was flawed only by the short but profoundly traumatic struggle between France and Prussia in 1870–1 (p. 1). In some respects prolonged peace saved lives but it also emphasised that material advantages might be enjoyed by the landowning peasantry if they limited their size of family and thereby held in check the division of property that had been spelled out in the Code Napoléon. For this and other reasons the 100 years after 1815 represent a critical demographic transition embracing a buoyant phase in the first half of the 19th century and a time of quite unparalleled stagnation in the second which was strikingly different from developments in surrounding countries.

Secondly, the period 1815–1914 begins with the kind of food crisis that had been typical of the ancien régime and reflected the limited possibility of moving food supplies rapidly before the railways were constructed (Jardin & Tudesq 1972). The crisis of 1816–17 affected the whole of Western Europe but was particularly grave in France not only due to climatic factors and epizootic disease but also because of the arrival of allied troops in 1815, whose presence disrupted harvesting in the most productive northern areas (Grantham 1978). The following year was cold and wet, with frost in May and widespread hail in August (Morineau 1976). Very poor harvests ensued, prices of corn and bread soared, and fear of starvation was expressed by widespread popular disturbance (Labrousse *et al.* 1970) (Fig. 1.1). In order to cope with this disaster, Russian wheat was imported through Marseilles for the first time but poor land communications meant that distribution beyond the Rhône valley raised formidable problems. Three decades later excessive summer drought in 1846 produced another crisis that affected not only cereals and other vital foodstuffs such as potatoes but also fodder supplies,

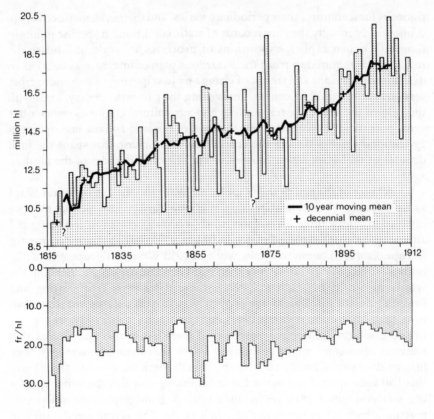

**Figure 1.1**   Wheat production and prices, 1815–1912.

which became desperately short in Normandy and Burgundy. Bands of beggars roamed the countryside and in many areas attempts by officials to redistribute the little food that was available were met with violent opposition, which was repressed with considerable severity. The following harvest mercifully proved much more satisfactory. Crises of such magnitude were not experienced again during the latter half of the century. Improved communications, in the form of the iron road and the iron-clad steamship, served to keep famine at bay by allowing food to be transported quickly between provinces and to be imported cheaply at times of hardship. The spectre of starvation vanished, but other components of French farming remained vulnerable to disaster after mid-century with, for example, a succession of vine diseases wreaking very real havoc in some parts of the country.

Last but by no means least, the choice of period is conditioned by the fact that many aspects of the changing mosaic of land use and agricultural activity in the years from Waterloo to World War I were quite well documented in both literary and statistical sources. A wealth of agricultural periodicals and economic journals was generated by national and local societies throughout the 19th century, with the *émigration intérieure* of legitimist landowners back to their estates in 1830 providing fresh stimulus for provincial agricultural associations. State concern for agricultural improvement and so-called interior colonisation during the Second Empire engendered a further flood of official and professional literature. Publications of this kind provide a valuable mass of empirical detail on the reality of change in the countryside but such evidence often defies quantification or effective summary. These deficiencies may be remedied by turning to the detailed statistics provided by the cadastral survey and successive agricultural enquiries.

## Facts and figures

In 1810 Napoléon I had inaugurated a system of national stocktaking so that a *Statistique générale* might be produced using evidence submitted by the prefect of each *département* (Bourgnet 1976). The attempt failed but a new project was announced in 1814 and a major agricultural survey was begun. Prefects were invited to submit observations on land use, crops and livestock to the Minister of the Interior and some administrators certainly complied, since fragments of information survive for 19 départements. These statements are of considerable interest and provide the first element in a broad sweep of fact finding about the French countryside that was to take place at fairly regular intervals during much of the 19th century (Festy 1956).[1] However, it soon became apparent that numerical estimates would have been more useful than impressionistic reports and in 1815 prefects were required to determine the prices and amounts of cereals produced and consumed in their départements that year (Newell 1973). The figures that they produced were no more than moderately informed estimates, since the land surface of administrative units was not known accurately, the metric system was poorly understood, and there was no mechanism whereby details of land use, cropping, productivity or prices could be collected at this time. The statistics that were submitted covered a limited number of topics and conveyed only the broadest impression of conditions at the département scale, but for all their weaknesses they complement the Statistique of 1814 and help portray many aspects of agricultural life in all parts of France at the end of the First Empire.

Annual returns were submitted in subsequent years but no major survey of farming was undertaken until the July Monarchy, following Louis-Philippe's announcement in 1835 that a new Statistique générale would be compiled (Clout 1980).[2] In the following year prefects were advised to ensure that agricultural information was gathered in every *commune* in their respective départements, and mayors were duly instructed to consult knowledgeable members of their community and then complete a standard questionnaire which dealt with conditions for 'the average year'. In an attempt to minimise errors, the Minister of Commerce recommended that committees be set up in each département to check, control and if necessary modify the mayors' returns. The first part of the questionnaire dealt with 24 crops or types of land use and required figures on surface occupied, total yield, average price, quantity used for seed, and the amount consumed in the commune. The second part sought to discover the numerical importance of 17 types of livestock and the average value of each category of beast. Information was required on 11 categories of livestock butchered each year, the weight of meat consumed and its average price. Many mayors viewed this enquiry with great suspicion, fearing that any information might be used for increasing taxes. Some gave literal rather than numerical replies, modified the questions that were specified, failed to use the metric system, or argued that the task was too great for them because of lack of knowledge or staff. Their returns were highly imperfect but hierarchies of revision committees, composed of 'enlightened men', were established for every *canton*, *arrondissement* and département with the task of correcting replies from individual communes. The revised returns were collated, summarised and eventually printed for each arrondissement and département in the land.

They represent the first nationwide survey of French farming to be undertaken simultaneously and expressed in metric measure. The range of information is broad but not exhaustive, since nothing was asked about the number or size of farms, tenurial arrangements, or the agricultural population and these omissions unquestionably limit the utility of the source. Emphasis was placed heavily on cereal production which reflected the administrators' obsession with food supply at this time of rapid population growth. It would be improper to claim that the département summaries are faultless, although it may be hoped that local excesses of over- or under-estimation would be averaged out by revision committees. One must admit that the Statistique was certainly not a rigorously precise portrayal of reality but one must acknowledge that in many ways it has to be accepted at face value. There is no comprehensive way of checking its contents since it had no precedent, nor may its information on land use be compared with that in the cadastral survey (*ancien cadastre*) since both classification and timing of compilation were different. There is no solution to this dilemma. Yet for all its shortcomings, the Statistique offers great interest and potential for the researcher. As a veritable mine of information on cropping and livestock early in the July Monarchy it may be used to establish a detailed data base that complements the more rigorous information on broad land-use realms in the ancien cadastre.

The agricultural enquiry of the July Monarchy (sometimes described as that of 1837) served as the model for later 'decennial' enquiries in 1852, 1862, 1882 and 1892 (Gandilhon 1969, Garrier 1967, 1977). Events of the Franco-Prussian War prevented an enquiry in 1872 but prefects were required to make brief returns relating to 1873 for an international survey. Methods of investigation for the decennial enquiries were similar to those employed in the 1830s, comprising detailed questionnaires to mayors and the use of committees for controlling results. The scope of questions broadened with successive enquiries to incorporate details of the rural economy, farm units and agricultural improvements, as well as land use, crops and livestock (Klatzmann 1955). Quality of information gradually improved, as the surface of each commune became known, use of the metric system became commonplace, and officials grew accustomed to reading questionnaires, making detailed investigations and answering the precise questions asked. Scope for human error and ill will in the communes remained enormous but control committees became more adept and provided reasonably accurate département summaries, although their labours never escaped criticism. The questionnaires were pruned of ambiguities and precise periods were specified for their compilation. Later enquiries required information on 'the average year' as well as the year in question, when the harvest may well have deviated from expectations. However, the introduction of such new questions, new categories and many new definitions means that information may not always be traced in a straightforward way from one survey to another. Some topics were virtually ignored until the final decades of the 19th century and are thus effectively excluded from analysis in the period 1815–1914; thus, for example, farm sizes were recorded partially in 1862 but not in full until 1882.

The final and most complete decennial enquiry before World War I was undertaken in 1892, with no successor until 1929. To some extent the 20-year gap at the end of our study period may be filled and the regular sequence of data continued by using the annual agricultural statistics, in particular those that were gathered in 1902 and 1912. The first detailed annual statistics had been collected in 1873 when over 100 questions on crop production and livestock were posed in each canton, controlled and summarised at département level. The method was less rigorous and the range of information narrower than in the decennial enquiries since nothing was asked about the rural economy, farm sizes or agricultural improvements, but the annual statistics offer formidable bodies of data nonetheless. While recognising their very real qualities, it would be foolhardy to assume that successive decennial and annual statistics could provide a rigorously accurate portrayal of agricultural reality. Nonetheless, these data were subjected to detailed scrutiny and 'correction' by administrators at several levels in the hierarchy and can be accepted as conveying the 'official' record of farming conditions for the period to which they relate. Even so, their detail and apparent precision must always be viewed critically and every figure cited in the present investigation should be read as if the word 'approximately' preceded it. Likewise, each message or interpretation to be advanced from such data should be treated with a similar measure of caution and qualification.[3]

The second major form of national stocktaking involved the vast cadastral survey that encompassed every plot of private and communal property in the land (Clout & Sutton 1969). With the fall of the ancien régime, old systems of taxation were abolished and replaced by a levy calculated on the amount and quality of property held by each landowner. Initial attempts to operate this reform equitably failed, since a detailed ground survey of plots throughout the country was needed. This task was not begun until 1807 and by 1814 only 9000 communes had been mapped. The fall of the First Empire retarded operations but work soon recommenced and by mid-century the whole of metropolitan France had been surveyed and documented. This ancien cadastre generated an enormous array of maps and registers which chronicled the ownership, use, quality and fiscal evaluation of each parcel.[4] Subsequent changes in ownership were noted, but not until 1907 was the register of landowners reconstructed afresh and a completely new statement on land use prepared. In addition to these sets of information produced near the beginning and end of the study period, two revisions of land revenues were prepared by tax officials following legislation in 1851 and 1879.[5] Methods of investigation were less rigorous than the exhaustive surveys of the ancien cadastre and the 1907 enquiry but both revisions contain valuable statistics on land use which may be set alongside data for the early 19th and early 20th centuries.

Records emanating from the cadastral surveys present both advantages and drawbacks for the researcher. An initial disadvantage is the timespan involved in drawing up the ancien cadastre, thereby precluding an exact cross section in time for the whole of France. However, the median survey date in each département is known and so adjustments may be made when calculating aspects of change between the ancien cadastre and the mid-century revision. A second complication derives from the fiscal rather than the agricultural objective of the cadastral system. All information relates to landownership units and no data are included on farms, crops, yields or livestock; such matters may only be elucidated from the agricultural enquiries. A third problem stems from the fiscal nature of the land-use classification, which was compounded by the surveyors' use of local terminology to describe some categories in commune registers. Fortunately this difficulty was overcome by an official reclassification of land-use details into five major realms (arable, permanent grass, vines, wood, waste) for each département. A fourth complication arises from the fact that state-owned land (usually under timber) was not subject to land tax and was therefore excluded from the cadastral statistics, but alternative sets of data can be introduced to cope with this omission.

A number of advantages counterbalance, and indeed in my opinion they outweigh, these limitations. Precision of ground survey and recording was far superior in the ancien cadastre than in early agricultural enquiries. The cadastral survey was conducted for fiscal purposes by its own staff and did not depend on the varying intelligence, literacy and good will of local mayors. Cadastral documents provide information at a range of scales, from the most intense detail of the land parcel to the manageable generalisation of départements which will be used throughout the present study. It is therefore possible to reconstruct the five land-use realms for four chronological cross sections: the early 19th century, mid-century, the late 1870s, and close to the outbreak of World War I. The statistical importance of each realm may be compared through successive cross sections in order to calculate net changes in land use during the intervening phases. Unfortunately, precise dynamics may not be demonstrated in this way, they may only be inferred. Detailed reconstructions of land use for the early 19th and early 20th centuries have been compiled by the late Aimé Perpillou (1977) and by other scholars but no serious attention has been paid to the revisions of 1851 or 1879 (since commune statistics were rarely conserved) nor have attempts been made to determine directions and rates of change through time.[6]

## Ways and means

The chronicle of land use and agricultural production in France from the fall of the First Empire to the eve of World War I is recorded in these vast and impressive surveys, each of which contains information at the département scale (Fig. 1.2). Cadastral surveys and revisions furnish land-use information of a high degree of accuracy which spans three phases (I – Restoration and July Monarchy; II – Second Empire and early Third Republic; III – 1879–1907). The agricultural enquiries contain variable, more frequent but rather less accurate information at a number of points in time between 1814 and 1913 (Augé-Laribé 1945, Klatzmann 1961). By virtue of their temporal specificity the data that they contain reflect the impact of short-term climatic variations as well as broader economic developments or technological changes, with crop yields being particularly sensitive to physical influences such as these. The century between Waterloo and World War I commenced with a series of cold springs and summers which extended until 1820, with a further span of cold years being encountered in the late 1840s and early 1850s (Le Roy Ladurie 1972). By contrast, the period from 1857 to 1875 was characterised by a succession of warm and clear (or at least average) springs and summers which generated a run of extremely productive agricultural seasons. During the 1880s conditions tended to worsen but started to ameliorate once again in the final decade of the 19th century. In addition, individual seasons displayed significant deviations from what might be expected during these climatic pulsations. Calculation of average values over a short span of years and inclusion of official statistics for the 'average year' provide a means of coping with the unique conditions of individual years but not, of course, with the implications of climatic trends that lasted for perhaps a decade or even more.

**Figure 1.2**   Départements.

Both agricultural and cadastral data display points of weakness as well as strength and are particularly unsatisfactory with regard to contextual variables such as farm size, land tenure, rotations, labour inputs and environmental parameters (Pautard 1965). Such deficiencies not only prevent the computation of statistical correlations that would be both meaningful in space and faithful through time but also preclude the introduction of numerous themes relating to the peasantry. Unfortunately farmland and farming people were investigated and recorded in separate and unrelated ways by officials during all but the final two decades of the 19th century. Nonetheless it may be argued that, when employed judiciously, the evidence of cadastral data and agricultural enquiries may be manipulated to enhance one another and thereby shed significant light on many aspects of the slow transformation of the French countryside between 1815 and 1914. As the title of the present study indicates, emphasis will be placed deliberately on the land of France and will in fact concentrate on its changing use and productivity. Exigencies of space mean that livestock husbandry and many contextual matters which enabled modifications in land use to occur, will receive only brief mention. Such matters, after all, have been explored by many writers and only the most salient points to emerge from their analyses will be recalled here.[7] Instead, a search for patterns of arrangements will be paramount in the present study and heavy reliance will be placed on quantitative cartography as a synthesising mechanism.

By way of introduction, the nature of land use and agricultural activity will be reconstructed at the end of the First Empire and attention will then be drawn to major forces that conspired to provoke changes in the use of the land of France during the following 100 years. The greater part of the subsequent analysis will be structured around the evidence of the cadastral surveys, with changes that may be inferred within and between land-use realms being examined in turn. Information from successive agricultural enquiries provides a finer scale of detail on arable cropping, agricultural improvements, livestock husbandry and other practices which will be inserted to enhance the record of the cadastre.

Considerable reliance must of necessity be placed on the comparison of evidence of static cross sections and the inference of intervening change. The hazards of this method are indeed appreciated, as is the possibility that actual processes at work between successive cross sections may have been infinitely more complex than direct comparison might suggest. But it must be stressed that in the absence of explicit evidence on process *per se* there is really no alternative to this approach. In addition, the sheer volume of statistical data to be reviewed necessitates the use of summarising devices to depict directions and average rates of inferred change over specified periods. This, too, is undertaken in full recognition of the possibility that changes over shorter spans of years may have been considerably more intricate. Finally, it must be understood that the département forms the most detailed spatial unit for which data are available consistently between 1815 and 1914; hence these administrative areas are employed throughout the analysis, although it is recognised that many départements display profound internal diversity (Fig. 1.3). To operate at a finer spatial scale is simply not possible.

In order to cope with the changes in the territorial extent of France in the 19th century it has been decided for purposes of this analysis to retain only the 82 départements that remained in French hands between the fall of the First Empire and the outbreak of World War I. Thus, the Alpine départements (Alpes-Maritimes, Savoie, Haute-Savoie) that became French after the Treaty of Turin in 1860 are not included, nor are the northeastern départements (Bas-Rhin, Haut-Rhin, Moselle) that were lost following the

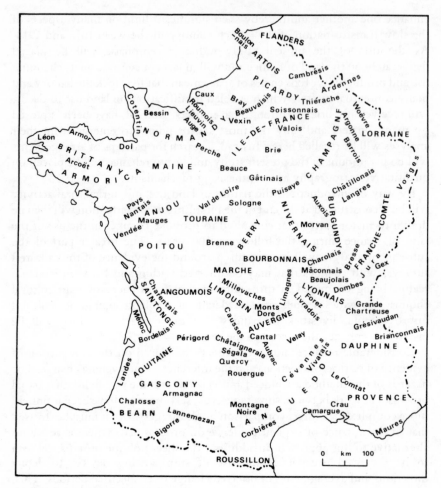

**Figure 1.3** Provinces and pays.

Treaty of Frankfurt in 1871. Corsica is also excluded by virtue of the unreliability and inconsistency of the island's statistical record. The spatial framework used in the following chapters to depict data on quantitative maps is the département outline of the final years of the 19th century. Terms such as 'France', 'the nation' or 'the country' are used throughout to describe the 82 départements rather than any larger area.

*Notes*

[1] The official agricultural enquiries and statistics that are used in this study are listed below, in chronological order. By virtue of the fact that each source relates to a readily identified year or

span of years it has not been felt necessary to repeat the full title of statistical sources in the text. In any case, exigencies of space would make it impossible to make reference to specific pages from which figures are drawn.

Ministère de l'Instruction Publique et des Beaux-Arts, Comité des Travaux Historiques et Scientifiques, Section d'Histoire Moderne (depuis 1715) et d'Histoire Contemporaine 1914. *Notices, inventaires et documents: la statistique agricole de 1814.* Paris: Rieder.

Ministère des Travaux Publics, de l'Agriculture et du Commerce 1837. *Archives statistiques.* Paris: Imprimerie Royale.

Ministère des Travaux Publics, de l'Agriculture et du Commerce 1837. *Statistique de la France: territoire et population.* Paris: Imprimerie Royale.

Ministère des Travaux Publics, de l'Agriculture et du Commerce 1840–2. *Statistique de la France: agriculture.* 4 vols. Paris: Imprimerie Royale.

Ministère de l'Agriculture, du Commerce et des Travaux Publics 1858. *Statistique de la France: statistique agricole 1852.* Paris: Imprimerie Impériale.

Ministère de l'Agriculture, du Commerce et des Travaux Publics 1868. *Statistique de la France. Agriculture: résultats généraux de l'enquête décennale de 1862.* Strasbourg: Berger-Levrault.

Ministère de l'Agriculture, du Commerce et des Travaux Publics 1868. *Enquête agricole.* Paris: Imprimerie Impériale.

Ministère de l'Agriculture, du Commerce et des Travaux Publics 1869. *Enquête agricole.* Paris: Imprimerie Impériale.

Statistique Générale de la France 1876. *Statistique internationale agricole 1873.* Nancy: Berger-Levrault.

Ministère de l'Agriculture 1878. *Récoltes des céréales et des pommes de terre de 1815 à 1876 (relevé des rapports transmis annuellement par les préfets au Ministère de l'Agriculture et du Commerce.* Paris: Imprimerie Nationale.

Ministère de l'Agriculture 1887. *Statistique agricole de la France: résultats généraux de l'enquête décennale de 1882.* Nancy: Berger-Levrault.

Ministère de l'Agriculture 1887. *Album de statistique agricole: résultats généraux de l'enquête décennale de 1882.* Nancy: Berger-Levrault.

Ministère de l'Agriculture 1897. *Statistique agricole de la France: résultats généraux de l'enquête décennale de 1892.* Paris: Imprimerie Nationale.

Ministère de l'Agriculture 1903. *Statistique agricole annuelle 1902.* Paris: Imprimerie Nationale.

Ministère de l'Agriculture 1913. *Statistique agricole annuelle 1912.* Paris: Imprimerie Nationale.

Ministère de l'Agriculture 1914. *Statistique agricole annuelle 1913.* Paris: Imprimerie Nationale.

[2] My earlier volume provides a fuller criticism of the Statistique Générale and, indeed, a longer, more detailed and much more explicit discussion of agricultural conditions throughout France during the 1830s.

[3] All absolute figures derived from agricultural and cadastral sources have been rounded to the nearest five or ten. Computed percentages normally have been summarised to the first decimal place.

[4] Département statistics from the ancien cadastre and subsequent revisions have been drawn from the following official sources:

Ministère des Finances: Direction Générale des Contributions Directes 1884. *Nouvelle évaluation du revenu foncier des propriétés non-bâties de la France, faite par l'administration des contributions directes en exécution de l'article premier de la loi du 9 août 1879: tableaux graphiques; atlas.* Paris: Imprimerie Nationale.

Ministère des Finances: Direction Générale des Contributions Directes 1913. *Evaluation des propriétés non-bâties; loi du 31 décembre 1907. Rapport de Monsieur Charles Dumont, Ministre des Finances, sur l'ensemble des opérations.* Paris: Imprimerie Nationale.

[5] Collection of new information undoubtedly involved a span of months or even years following the authorising legislation for these revisions but, for sake of convenience, cadastral statistics deriving from them will be identified by the years 1851, 1879 and 1907.

[6] Professor Perpillou produced over a score of articles which contain detailed cartographic reconstructions of cadastral land use for various regions of France. These works were summarised in the form of three national maps at the scale of 1 : 1 000 000. A valuable overview of the complete project has been provided by Solle (1981).

[7] Recent publications which probe and synthesise economic and social changes in 19th-century France include Barral (1979), Duby and Wallon (1976), Houssel (1976), Laurent (1976), Price (1981), Weber (1977) and Zeldin (1973, 1977). Each provides valuable contextual information for the present investigation and several raise challenging ideas and insights.

# 2  Legacies of empire

## Land and people

In many parts of France the quarter century from the Revolution of 1789 to
Napoléon's defeat at Waterloo proved a time of profound upheaval and
disruption to everyday life. In some ways these years saw the emergence of a
'new' France since the feudal order had been abolished, new codes of law and
landownership had been fashioned, and new systems of administration and
taxation had been created, but in more fundamental respects little had
changed (Soboul 1968). With its great variety of physical and human resources France remained exactly the same summary of Europe that it had
been at the close of the ancien régime. Its component pays were as diverse as
ever; their inhabitants observed local customs, spoke patois and retained
traditional loyalties to ancient provinces, while contrasting ways of life in
the Midi and Nord continued much as before (Demonet et al. 1976, Le Bras
& Todd 1981, Prince 1977). The nation's economic unity, that had been
proclaimed at the Revolution, was more promise than reality and was to
remain that way until communications were transformed during the course
of the 19th century (Fox 1971).

Almost all Frenchmen were countryfolk and the majority of these were
food producers, ordering their lives by the farming calendar and praying
fervently that they might grow enough food to survive, with perhaps a little
to spare. In spite of important examples of internal migration, most families
lived essentially local lives and rarely ventured beyond their village or nearest
market. Only the very few, who lived near a town or major routeway or
commanded particularly distinctive agricultural resources, geared part of
their activity to producing food for townsfolk (Joigneux 1847). All were
vulnerable to the vagaries of the elements, with a bad year or the failure of a
particular crop causing food shortages and perhaps raising fears of
starvation. Emergency supplies, even if they were to be had, simply could
not be moved with sufficient speed down muddy tracks and rutted roads,
along torrential rivers and shallow canals, or across dissected terrain, to
prevent hunger taking its grip and the spectre of famine haunting town and
country alike. The fragility of survival was rendered even more precarious
by war or civil disturbance, with recruitment of troops, requisitioning of
horses and commandeering of supplies, if not actual fighting, making
agricultural activities difficult in the extreme (Rollet 1970). Disruption
associated with allied invasion and occupation from January 1814 to
November 1818 was compounded by poor harvests and serious food
shortages in 1816 and 1817 which provoked a temporary break in the
generally buoyant demographic climate.

The population of the whole of France had increased by some 7 million during the 18th century, with 1 million of that total being added by the annexation of Lorraine and Corsica. The tempo of growth had accelerated after 1750 and, despite the occurrence of individual 'hollow' years, a total of 25 906 350 was recorded in 1801 for the 82 départements. When the next reliable figures became available 20 years later, the total had risen by 11.3 per cent to 28 856 600. Such demographic development clearly provided a very positive stimulus to enhance agricultural production (Goy & Le Roy Ladurie 1982). At this stage censuses did not provide a definition of rural and urban areas but it is possible to separate the residents of *chefs-lieux d'arrondissement* from their country cousins. Some 3 708 840 people lived in chefs-lieux in 1801, representing 14.3 per cent of the national population; 20 years later the same administrative towns housed 4 153 350 (up 12 per cent) but their significance in the national total remained virtually unaltered (14.4 per cent) (Fig. 2.1a). In fact, three-quarters of the chefs-lieux contained fewer than 10 000 people apiece and many small towns were little more than villages writ large. Twenty-three départements in eastern France, the Alps, and Massif Central had no town of more than 10 000 people in 1821, but the nation also contained 63 towns with populations of 10 000–25 000 (housing 995 000), 17 with 25 000–50 000 (housing 537 900) and eight even greater cities (Fig. 2.1b). Above-average measures of urbanisation were usually produced by large, free-standing cities, although in the Mediterranean départements and Nord there were important clusters of sizeable towns. Only Seine (88 per cent), Bouches-du-Rhône (48 per cent) and Rhône (40 per cent) had over two-fifths of their population resident in chefs-lieux d'arrondissement in 1821 (Dupeux 1974).

As the centrepiece of administration, located astride a major system of navigation, set at the focus of the nation's highways and in the midst of fertile cornlands, Paris had grown phenomenally of recent decades, with its population rising from an estimated 524 000 in 1789 to 546 850 in 1801, 622 600 in 1811 and 713 960 in 1821. During the first two decades of the new century the capital had increased its citizens by more than the total population of Lyons, the second city in the land. With an estimated population of some 670 000 at the fall of the First Empire, Paris was truly exceptional, commanding much larger tributary areas for food and for migrants than any other French city. In addition, it required vast supplies of wood for domestic and industrial purposes, which amounted to 1 200 000 m³ (1.80 m³ per inhabitant) in 1815 (Huffel 1904). Lyons housed 149 170 in 1821 and had grown by 39 670 (roughly the population of Nîmes) in the preceding 20 years, but the next three cities in the urban hierarchy (Marseilles 109 480, Bordeaux 89 200, Rouen 86 730) each contained rather fewer inhabitants than at the beginning of the century. Indeed, no less than eight départements appear to have experienced a decline in their number of town dwellers between 1801 and 1821.

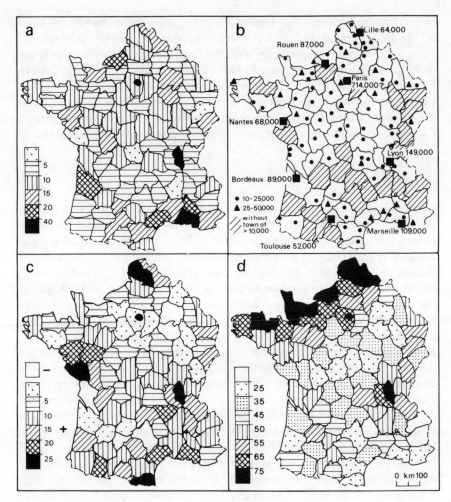

**Figure 2.1** (a) Urban population, 1821 (per cent living in chefs-lieux d'arrondissement); (b) major towns, 1821; (c) change in rural population, 1801–21 (per cent); (d) density of rural population, 1821 (/km²).

Rural France, lying beyond the chefs-lieux, increased its population by 11.2 per cent during those two decades, with particularly rapid growth affecting not only the environs of great cities like Paris, Lyons, Marseilles, Montpellier and Nantes, but also occurring in remote country areas like Haute-Loire and Hautes-Pyrénées and the pacified territory of Vendée (Fig. 2.1c). In complete contrast, five départements in Champagne and the south-west would appear to have experienced precocious depopulation. On average, the French countryside supported 50 inhabitants/km² in 1821 through working the land and engaging in a wide and intricate range of crafts, by-employments and service activities. Rural areas close to the heart of the Ile-de-France and in Lyonnais were densely settled, as might be expected, but more surprisingly were the high densities of rural population in a dozen départements close to the Channel coast (Fig. 2.1d). Eastern France, the lower Loire, middle Garonne and the pays charentais all supported above average densities. The harsh environments of the Alps and the southern Massif Central were relatively empty lands and middle France, Champagne and much of Provence supported fewer than 35 persons/km².

## The arable mosaic

Grain growing was unquestionably the pilot sector of the rural economy at the end of the First Empire, with no less than half of France being under the plough according to the ancien cadastre. In fact these returns were not completed until rather later in the century and therefore convey a somewhat inflated picture for 1815. In addition, high cereal prices of 1812 and 1816–17, significant increases in population and the survival of basic technologies in the countryside all encouraged wasteland clearance and extension of crop-land in the first half of the 19th century rather than widespread ex-perimentation to raise the productivity of existing fields (Bergeron 1972). More land was being tamed at this time in many pays, such as Rouergue where défrichement following the Revolution had already allowed much more food to be grown (Monteil 1803). Highly unstable environments were also cleared, as in Ardèche where the conseil général recorded substantial deterioration of newly cropped hillsides, and in the mountains of Auvergne where Yvart (1819) claimed a thousand examples of sloping expanses of soil being cleared, cropped and then washed away (Bozon 1961). Arable land exceeded three-quarters of the surface of the great plateaux of Artois, Picardy and Champagne, while four-fifths of Beauce was under the plough (Fig. 2.2a). Proportions were substantially lower in northeastern France, Lower Normandy and especially to the south of the Loire, where arable covered more than 60 per cent of the surface only in Bourbonnais and the middle Garonne. Combinations of environmental resources and densities of population in many southern areas were clearly quite different from those further north, with many southern départements having less than half of their land under the plough.

The use of land within the arable realm was orchestrated by the particular crop rotations that were followed, through deference to tradition or because of stipulation by lease, since crop yields would fall unless the soil were rested or fertilised adequately, while continuous cropping for the same product tended to depress output and increase vulnerability to plant disease (Lullin de Châteauvieux 1837–8). Equal alternation of cropping and fallowing (when land would be worked before seedtime) provided the simplest solution to

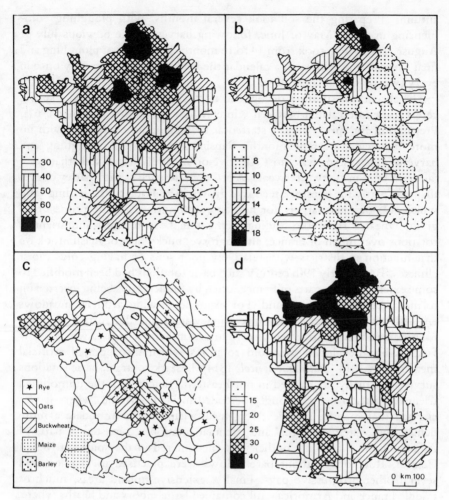

**Figure 2.2** (a) Arable as percentage of total land use according to ancien cadastre; (b) average wheat yields 1815/20 (hl/ha); (c) dominance of cereals other than wheat, 1815; (d) value of production according to Chaptal (fr/ha).

such matters and, in the early 19th century, biennial rotations were practised to the south of a line from the Gironde, skirting north of the Massif Central to Lake Geneva, and also in Vexin, Vendée and around the lower Loire (Musset 1952, Sigaut 1976). The ecology of the Midi, in particular its summer aridity, restricted the range of cereals that might be grown successfully (Bénévent 1938). Hence spring grains were rarely sown since they would have a precarious existence in a growing season cut short by summer heat and drought. Instead, it was better to concentrate on winter corn, particularly wheat and to a lesser extent rye, which could be sown in autumn. Preparing the soil took several months, with ploughing commencing in April, May or June, following harvest in the previous July or August. The system took from 14 to 16 months between first ploughing and final harvest, and with such a calendar the land could be sown only once in every two years (Faucher 1934).

The absence of summer drought in northern France permitted a spring cereal to be intercalated between winter wheat and fallow (Faucher 1961). Preparation for spring cereals started at the beginning of winter, with no more than eight or nine months elapsing between first ploughing and harvest. This system was most effective on naturally fertile soils that were fallowed once in every three years. A *petite céréale*, such as oats, barley or rye, that was partly for livestock feed was grown in addition to the leading crop which was destined largely for human consumption. As the prime bread grain, wheat formed the leading element in both biennial and triennial rotations over much of France, although rye, buckwheat or maslin took on that function in districts with relatively poor soils or having cold, moist climates. By the early 19th century such basic rotations had been modified to some extent in many parts of France, often by fodder crops being inserted on former sections of fallow land. For example, sowing artificial meadows enabled tenant farmers in Beauce to deviate from the formal triennial rotation and this practice was also followed in Seine-et-Oise, where a few farmers had developed four-fold rotations which incorporated artificial meadows by 1814 (*Statistique agricole* 1814). In stark contrast, basic rotations survived virtually unchanged in northeastern France, with the unmodified triennial rotation being strongly defended in Haute-Marne, where it was claimed that manure supplies were insufficient to accommodate a more demanding system. Biennial and triennial rotations co-existed in parts of Aquitaine and a triennial outlier was encountered in the southern Massif Central. Rotations without bare fallows were·practised in Flanders and several other progressive pays. Finally, extensive upland areas, much of middle France and Armorica, still contained large moors and heaths where phases of cropping were interspersed with long fallows. Thus, in the interior of Finistère, buckwheat, rye or oats were produced for a couple of years and the soil then rested for a further five or six seasons; by contrast, the land was fertilised so heavily in nearby coastal districts that it could stand cropping every year.

This information helps to establish a framework within which detailed evidence on arable husbandry may be set. Wheat may be expected to occupy a particularly dominant position in the arable mosaic of southern areas where biennial rotations were practised, while proportions of one-third or rather less might suggest its presence in a triennial rotation. According to the prefects' estimates, wheat covered 4 463 800 ha in 1815 and was the most extensive cereal in 48 départements, especially in the south-west and in northern France. It flourished best in fertile, clayey soils and some strains grew successfully in less favoured environments, but really harsh climates and poor soils reduced its significance in the arable realm of mountainous regions and *mauvais pays* of the lowlands (Castellan 1962). By the end of the First Empire wheat cultivation was advancing substantially to replace more rustic cereals in many areas, with, for example, rye and oats retreating in Maine-et-Loire as wheat growing advanced. This trend was to become much more pronounced as the 19th century wore on.

Evidence from the July Monarchy implies that seeding practices must have varied greatly from pays to pays in the early 19th century but they were not in fact recorded in 1815 and hence such useful indicators as seed : yield ratios may not be calculated for such an early date. Annual gross yields fluctuated enormously in response to weather conditions at critical phases in the farming year and 1815–17 proved to be a very difficult time, with the national average for 1815/20 being only 9.88 hl/ha (hectolitres per hectare) and annual averages ranging from 8.59 hl in 1815 to 11.40 hl in 1818. Output varied strikingly between provinces, with gross yields for 1815/20 reaching almost twice the national average around the capital (Seine 18.85 hl) and in Flanders (Nord 18.10 hl), and remaining high in the Ile-de-France (Seine-et-Oise 16.95 hl), the northern Paris Basin (Pas-de-Calais 16.88 hl, Somme 15.80 hl, Oise 15.50 hl) and the fertile coastal belt of Finistère (15.44 hl) (Fig. 2.2b). Output fell below 8.00 hl/ha throughout middle and south-western France and was close to half the national average in Dordogne (4.37 hl) and Haute-Marne (4.70 hl).

As well as forming the second element of the triennial rotation in northern France, oats was successful in a wide range of other environments and required relatively little attention, unlike the ploughing and manuring that preceded the sowing of wheat. However, extremes of heat and drought reduced yields of oats and cold, wet conditions prevented the grain maturing, although green oats served as an acceptable fodder. Oats entered the human diet in some parts of northern France but its real importance was as a feed for horses. These facts help explain why oats was reported to cover 2 440 075 ha in 1815 and was more extensive than wheat in southern and eastern stretches of the Paris Basin and in parts of Aquitaine and Lower Brittany (Fig. 2.2c). A similar amount of land was devoted to rye (2 520 710 ha) as this crop flourished on acid soils and in many harsh environments, with 11 départements in the Massif Central devoting more land to rye than wheat and smaller *ségalas* being found in Maine, southern Brittany, Champagne and valleys in the Alps. Rye was sown in autumn and was robust enough to survive long periods of snow cover but was then at the mercy of spring frosts or excessive rain. As well as providing the grain for black bread, rye yielded useful quantities of straw and was used as a source of fodder in northern France, where it appeared sometimes as a petite céréale in the second year of the triennial rotation. Rye and wheat were sown together in varying proportions in the form of maslin, which was usually grown on moderate soils where the robust qualities of rye were appreciated. Maslin covered 902 300 ha in 1815 and was quite popular in northern France but in no département was it more extensive than wheat.

Buckwheat was suited to most kinds of light soils and did not require heavy application of manure. It needed only a short growing season with a reasonably high degree of humidity, hence hot drying winds or late frosts could cause serious damage. The crop was reported as covering 653 575 ha in 1815 and occupied more land than wheat in nine départements of Armorica and Limousin, where it yielded a useful grain for human subsistence, while in other parts of France it was given green to livestock or its grains were fed to poultry. Barley was grown on 1 007 310 ha in 1815, flourishing on calcareous soils on the fringes of the Paris Basin and in upland areas, where its short growing season was particularly appropriate. Its grain was consumed in breweries, stables and poultry yards and entered the human diet in parts of eastern France. In addition, it was cut green for feeding to horses; even so, only in Haute-Loire did it occupy more land than wheat. Maize covered the smallest area (537 975 ha) of the seven cereals recorded in 1815 and was concentrated in the south-west and the Rhône valley, since the strains being grown at that time required mean temperatures exceeding 18 °C between May and September. Maize suited many types of soil but needed heavy manuring and a great deal of hoeing if high yields were to be obtained. Such requirements were best satisfied in Basses-Pyrénées and Landes where it occupied more land than wheat.

Since other crops were not recorded by the prefects at the end of the First Empire, it is not possible to reconstruct the complete pattern of arable farming or to determine crop combinations in 1815. Two years later an estimated 500 000 ha were devoted to potatoes, with foci of cultivation in the north-east, the central Pyrenees, Maine, the eastern Massif Central, Limousin and Périgord. No statistics were gathered on the fodder crops (clover, lucerne and sainfoin) known collectively as 'artificial meadows', which occupied previously fallow sections of ploughland in the Paris Basin and in other fertile and relatively innovative parts of the country, where the task of generating ever greater quantities of cereals for the local population did not form the sole and virtually unqualified objective of agricultural activity (Chorley 1981). Many respondents to the Statistique of 1814 commented that artificial meadows were being sown more widely than at the time of the Revolution (Festy 1957). Nonetheless, drought hampered their cultivation in southern France, although they would respond well to irrigation. Traditionally minded farmers were reluctant to experiment with them, as in Tarn-et-Garonne where the majority of cultivators refused to acknowledge their advantages and continued to rely on stubble, straw and maize stalks as sources of fodder.

By contrast, artificial meadows had met with great success since 1800 on the plateaux of Côte-d'Or, especially in localities that lacked natural grassland, while around Troyes the rental value of natural meadows had actually declined following the introduction of clover and sainfoin. In the Ile-de-France the popularity of artificial meadows had increased greatly following the introduction of merino sheep on the Rambouillet estate in 1787. These new fodders were having an almost incalculable effect, allowing more live-stock to be kept and more manure to be returned to the soil to keep it in good heart. Some farmers near Rambouillet had sown a quarter or even a third of their ploughland with artificial meadows by 1814 and any downward fluctuation in cereal prices encouraged them to grow more. Only farmers who neglected to calculate their likely profits failed to devote more land to fodder crops in that particular district; but such a high degree of popularity was, of course, rare. In Nièvre too few artificial meadows were being grown to have any influence on local agriculture and in parts of Armorica they were still virtually unknown in 1814.

Many travellers and writers regretted that so much land was left as bare fallow each year or was cultivated for only a few seasons before being allowed to rest and revert to rough grazing. Indeed, bare fallowing was condemned by many as the major weakness of French farming at this time although it did, of course, enable large numbers of sheep to be grazed which provided dung to fertilise the soil. Little is known about the precise proportion of land left fallow at the close of the First Empire, since traditional rotations had been modified to some extent in many parts of France. Undoubtedly less land was left *en jachère* than the continuing operation of biennial and triennial systems might be taken to imply. Yet even in the progressive département of Seine-et-Oise many farmers clung on to fallowing since they feared that more intensive cropping would necessitate expenses that they would not be able to bear; while fallowing was judged to be indispensable in more arid environments, such as Tarn-et-Garonne where the biennial rotation reigned supreme. Neither the *enquête agricole* of 1814 nor the statistics of 1815–20 provide a means of summarising the productivity of French ploughland at this time, however Chaptal (1819) estimated that each hectare generated 28 fr worth of produce (Fig. 2.2d). The superiority of Seine (216 fr/ha), Nord (69 fr/ha) and Seine-Inférieure (67 fr/ha) was unquestionable, with eight other northern départements yielding over 40 fr/ha, while in parts of the Alps, Massif Central and middle France the figure plummeted below 15 fr (Coyaud 1974).

## Remaining realms

By contrast with the wide extent of arable land, less than one-tenth of France was under natural grassland at the fall of the First Empire. Well watered vales and claylands were devoted to meadows only if local pressures to produce cereals would allow (Castellan 1960). In any case, natural flooding often

reduced the quality of hay that could be produced, as in the lower valley of
the Vilaine. The most extensive grasslands were found in Lower Normandy,
Limousin and Velay, with no less than one-quarter of Haute-Vienne being
under grass (Fig. 2.3a). At the other extreme, riverside meadows in the
arable départements of the southern Paris Basin occupied only a tiny fraction
of the surface and failed to generate enough hay for local needs. Thus,
despite the widespread cultivation of artificial meadows in Seine-et-Oise
great quantities of fodder had to be carried into that département each year.
The amount of land devoted to natural grass was slight across much of

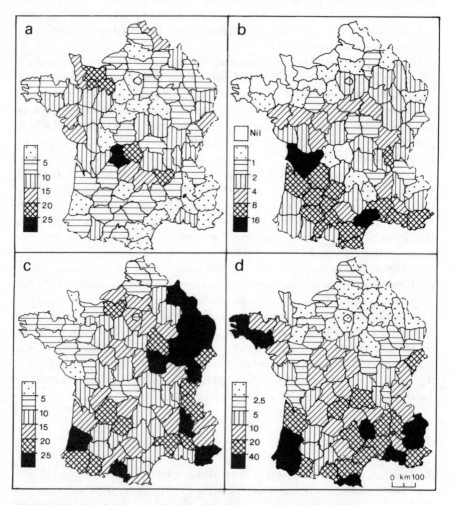

**Figure 2.3** Land use according to the ancien cadastre (per cent): (a) permanent
grass; (b) vines; (c) cadastral and state woodland; (d) wasteland.

southern France where there was insufficient water for irrigation, and even in areas where that technique was practised, such as the valleys of the southern Alps and the Pyrenees, there was the problem of expense and the ever-present threat of devastation as flood waters deposited material that had been eroded upstream. Meadow irrigation was much more easily accomplished in the more humid environments of Limousin and Lower Normandy, with four-fifths of the meadows around Vire being 'watered' (Garnier 1975).

Chaptal (1819) estimated that France supported 2 million horses, almost 7 million head of cattle and no fewer than 35 million sheep at this time. Very often their quality was poor and in many regions they only managed to survive because of the resources of the wastelands and the arable realm, which afforded fallow and stubble for grazing as well as fodders and cereals for feeding direct to livestock (Crubellier 1975, Fiétier 1977). The practice of stubble grazing (*vaine pâture*) provided fodder for sheep and was widespread in much of the Paris Basin and Aquitaine, but in more progressive areas, such as Seine-et-Oise, it was criticised by good farmers who insisted that it hindered the introduction of new rotations. They also wished to abolish fallowing so that more land might be devoted to growing fodder. The inter-commoning (*parcours*) of sheep from neighbouring villages across harvested openfields continued to operate over much of northeastern France, as in Haute-Marne where parcours and vaine pâture had been in use from time immemorial. However, both forms of collective grazing were in decline in parts of the south and were little used in areas such as Nièvre where much ploughland was already enclosed but wide areas of uncultivated land still survived to provide abundant rough grazing (Thuillier 1956).

Just under one-twentieth of France was under the vine at the time of the ancien cadastre, with viticulture only being absent from Creuse and nine départements close to the Channel (Fig. 2.3b). Wine was not an everyday beverage for humble folk in such areas, although it might be consumed at times of illness (Corbin 1975). Many sunny hillslopes close to the northern limit of cultivation carried vines as in the département of Seine-et-Marne, although Brie wines were condemned for being 'hard and cold' (Tresse 1803, p. 88). Freedom to cultivate one's land as one wished had been proclaimed at the Revolution and subsequently viticulture increased apace by virtue of the chance of profit that it offered. Greater quantities of high quality and ordinary wines were produced, although as a general rule the quality of wine tended to fall as more land was planted up (Bergeron 1972). Modifications in transportation produced important reactions in the wine trade with, for example, the opening of the Ourcq canal in 1808 linking the rivers Seine and Marne and allowing the wines of Burgundy to reach Paris. This, in turn, accelerated the decline of local viticulture in some parts of the Ile-de-France (Ackerman 1978). Closure of the eastern frontier in 1814 prevented Lorraine wines being exported down the Moselle and the Rhine, and this provoked an increase in production of modest wines to satisfy growing local consumption

(Parisse 1978). The cadastral figures undoubtedly provide an overestimate of viticulture in 1815, with 1 550 000 ha perhaps being a more realistic total (see Ch. 8). Nonetheless, the broad pattern derived from the cadastre was valid for earlier years even though it must be appreciated that the département scale is really too coarse to display the specific environmental requirements of the vine. This land-use realm occupied four times the national average proportion of land in Lower Languedoc and the pays charentais, and more than double the average in Lyonnais and much of the Midi and south-west; while vineyards along the valleys of the Loire, Rhône and Saône stood out as important local concentrations.

Woodland covered one-sixth of France at the time of the ancien cadastre and it is likely that the proportion had been slightly higher in 1815 since déboisement continued vigorously during the early decades of the century. According to the cadastre, no less than 13 départements had over one-quarter of their land under trees, with above average proportions throughout eastern France and figures exceeding 30 per cent in Nièvre and Var (Fig. 2.3c). Quality was often poor following the Revolutionary period and the Annales de Statistique (1802) contain numerous examples of recent devastation and deforestation. It was reported that two-thirds of the timber in Doubs had been felled in just over a decade and much rough grazing had been burned off and cleared (Anon., Doubs 1802). In Drôme, further south, 'all the wood-land had been devastated' and the few trees that remained were only there 'because of the idleness of the woodcutters or the shortage of labour to destroy even more trees' (Anon., Drôme 1802, p. 395).

Deforestation served to accelerate soil erosion so that previously manageable rivers were turning into rushing torrents which caused immense damage, as in the Tech valley of Pyrénées-Orientales (Tessier 1818). Likewise, in 1819 Prefect Dugied of Hautes-Alpes quoted the recollections of old inhabitants to help support his assertion that 'we have seen more fields and pastures swept away since 1789 than during the preceding two cen-turies' (cited in Huffel 1904, p. 129). He believed that half the land of the département had been rendered unproductive as a consequence of de-forestation and erosion, and urged that further clearance be forbidden and reforestation commenced, with the government purchasing and restoring the most devastated districts. Numerous respondents to the enquête of 1814 lamented the excesses of recent years in many parts of France; in upper Beaujolais much woodland had been felled to supply Lyons and surrounding towns with fuel and construction timber; while in Drôme fragile slopes had been deforested since 1789 and erosion, gulleying and deposition had taken place so that much low-lying farmland had to be abandoned as a result. Precisely the same kind of phenomenon had occurred in Aveyron where inconsiderate and excessive felling provoked severe timber shortages (Monteil 1803). However, chestnut trees covered roughly 500 000 ha in 1815 and provided an important subsistence crop in Périgord, the Cévennes and other hungry pays along the southern fringes of the Massif Central.

One-sixth of France was composed of marshes, moors, heaths and other forms of uncultivated land at the time of the cadastre and such forms of land use were particularly extensive across many of the southern provinces (Auvergne, Languedoc, Dauphiné, Provence), the Landes, the mauvais pays of middle France and Armorica (Fig. 2.3d). Much of this was commonland, as in the Landes where large areas were flooded in winter 'to resemble a great lake' and in Ille-et-Vilaine where heaths and marshes covered more than one-fifth of the département (Anon. 1823, p. 40; Anon. 1803, p. 145). According to Malte-Brun (1833), 'should the traveller leave the banks of the Loire . . . he may observe large heaths and desert plains' across much of middle France, and certainly the communal pâtureaux of Berry and Nivernais supported such meagre vegetation that it 'simply prevented livestock dying rather than helping them to live' (Vidalenc 1970, p. 97; De Chambray 1834). The topsoil of the Champagne pouilleuse was reported to be less than 2.5 cm thick but it provided rough grazing for numerous flocks of sheep and might even support timber if sufficient wealthy men could be found to invest in the area (Statistique agricole 1814). Uncultivated land in the far south yielded lavender, thyme and other herbs as well as a little grass and kindling wood. Everywhere the wastelands had their value in the traditional economy and were often strongly defended by commoners who wished to preserve their age-old ways (Papy 1947–8).

Reclamation of wasteland since 1789 had permitted the arable surface to increase in many pays, as around Sègre in Mayenne and along the Paris–Lyons highway in the Morvan where much land had been cleared and cropped successfully with rye. New farms had been established on gâtines near Cosne and Nevers and it was suggested that even more défrichement might be possible if favourable leases were to be granted. But some prefects were more cautious in their opinion submitted to the enquête of 1814. For example, in Deux-Sèvres new land was being put under the plough each year although it was argued that it might be better to farm existing arable more efficiently. In Tarn privately owned rough grazing was being cleared for rye, oats and potatoes but the land was really so poor that it could support only a couple of harvests before having to be abandoned to scrub. Excessive clearance in Tarn-et-Garonne had created a shortage of pasture land, and in the Alps and Pyrenees that problem was compounded with accelerated soil erosion so that several prefects urged that further défrichement be forbidden. The folly of excessive clearance was also realised in Ille-et-Vilaine, since adequate fertilisers were not available for freshly reclaimed land, and it was believed that much of the uncultivated soil was really too poor for reasonable crops to be harvested and clearing expenses to be recouped thereby. The defenders of the old order had common sense on their side but, as population numbers mounted and land hunger grew prior to mid-century, so the process of défrichement came to change the face of ever wider areas of land.

The mosaic of land use was undoubtedly being transformed in other important ways during the early decades of the 19th century. Some pays with good systems of communication by land or, more probably, by water were turning to modes of production that were geared emphatically to the needs of the market. Such was certainly the case across parts of the Paris Basin and in various individual districts in other regions of France, where some farmers were turning to precocious specialisation in viticulture or livestock husbandry. But these were exceptions to the general rule of the times; poor communications, simple technology, limited penetration of market forces and survival of communal traditions and practices (irrespective of the liberties proclaimed at the Revolution) meant that in 1815 by far the greater part of the French countryside continued to function much as it had always done. The majority of French families struggled to produce enough food to survive. Patches of moorland and the margins of forests came under continued attack, since extending the arable surface was a simple operation that involved no new forms of organisation or technology. Human muscle was all that was needed and that was in plentiful, indeed growing, supply. To intensify production from the existing farmed area surface was a far more demanding procedure which required new forms of fertilising, cropping and managing the land. The means of achieving such an objective lay well beyond the scope of most peasant households at this time, with few resources at their command save their land and physical labour. Extension was the traditional solution, intensification the radical innovation. In fact, both processes were to operate between 1815 and 1914 but in ways that were to prove strikingly different and which were to fluctuate markedly through time and across space.

# 3   Underlying forces

## Growth and redistribution

In the early years of the 19th century, France remained almost as com-
partmentalised spatially as it had been during the ancien régime. The
economy was dominated by food production and artisanal forms of
manufacturing; modes of transport were slow and expensive, which
hindered the movement of goods and people and the transmission of ideas.
Traditional *mentalités* prevailed, while localism and isolation survived save
along a few well maintained routes and navigable waterways. Construction
of canals and highways had certainly afforded some modest changes but it
was not until the building of railways was started during the July Monarchy
that passengers and commodities could be moved at speeds that surpassed
animal power. Locomotion along fixed rails was to liberate bulk movement
from many of the uncertainties that had surrounded navigation and road
transport in the past and, thereby, was to introduce a previously unknown
quality as well as a new dimension to land communication. Town and
country were gradually to become interlinked with an intensity that had been
inconceivable during the ancien régime économique, thereby allowing nodal
regions and their hinterlands to emerge even more powerfully. This new
form of *connectivité* was to enable the needs of the market to overtake the
harsh dictates of subsistence across an ever-widening spread of territory,
within which polyculture and self sufficiency were to be replaced by
specialisation and commercial farming. Without doubt it would be an
exaggeration to attribute such a transformation simply to the railways but
they were 'the major factor for change . . . enabling a balancing of supply and
demand and the elimination of traditional crises' (Price 1981, p. 84).

Rural responses to the opportunities offered by the new technology were
normally far from immediate since the burden of history continued to weigh
heavy in the countryside. Peasant landownership remained a remarkably
tenacious feature of French life and, for this reason, rates of urbanisation and
population growth after 1850 were slow by comparison with those in
neighbouring countries (Sargent 1958). Yet despite the disruptions that
followed the Revolution and occurred during the Napoleonic Wars, the
population of the 82 départements had risen from 25 906 350 in 1801 to
28 854 600 in 1821, and a comparable volume of increase had been achieved in
the next 15 years, reaching 31 850 000 in 1836. The tempo had declined
during the 1840s to yield a total of 34 005 650 in 1851, and growth was both
modest and uneven, reaching 38 454 600 in 1911. In short, France supported
one-third more inhabitants on the eve of World War I than 100 years earlier.

The distribution of that slowly growing population had changed profoundly during the 19th century, partly in response to spatial variations in natural change but more especially because of the operation of internal migration (Pitié 1971). During the July Monarchy 30 départements managed to attract more migrants than they dispatched but that figure declined thereafter, with three-quarters of the départements of France losing more migrants than they received during the years between 1851 and 1911 (Fig. 3.1).[1]

Paris was the leading destination for internal migrants throughout the 19th century and départements which contained Lyons, Marseilles, Bordeaux, Montpellier and the large cities of Flanders, Provence and Lorraine formed powerful points of attraction for part if not all of the period between 1836 and 1911 (Chevalier 1950). Net outmigration became the fate of ever wider sections of the French countryside with myriad reasons stimulating folk to seek their livelihood elsewhere (Armengaud 1951). Only nine départements lost the equivalent of more than 5 per cent of their 1836 population through net outmigration in the subsequent 15 years but during much of the second half of the century the figure was close to 20, and no fewer than 33 départements experienced strong net outmigration between 1891 and 1911 (Fig. 3.1). Seine département received a net inflow of migrants between 1836 and 1911 which surpassed its total population in 1836, and Bouches-du-Rhône, Rhône and Seine-et-Oise attracted volumes of net immigration that exceeded half their initial numbers. At the other extreme, 60 départements lost more migrants than they gained, with net losses being greater than 30 per cent of the initial population in 10 départements located mainly in the Massif Central.

The picture is rather less harsh when trends of natural change are combined with those of net migration. Nonetheless, exactly half of the départements in France contained fewer inhabitants in 1911 than in the early 19th century (Fig. 3.1f). Much of Normandy, Maine, the south-west, Massif Central, Alps and north-east made up vast regions of decline, with net losses of more than 20 per cent and even 30 per cent affecting parts of Gascony and Lower Normandy, Haute-Saône and Basses-Alpes (Price 1975). Départements containing dynamic urban centres developed quite differently, with Seine and Bouches-du-Rhône more than doubling their population between 1821 and 1911 and increases of more than 50 per cent occurring in Rhône, Loire, Seine-et-Oise, Pas-de-Calais and Nord (Carrière & Pinchemel 1963). As a result, the human geography of the French nation in 1911 was more profoundly centralised than it had been 100 years earlier. An official definition of 'urbanisation' had been used for the first time in the census of 1846, with 24.4 per cent of the French people living in 'urban' communes (i.e. having over 2000 in their largest settlement) and 17.5 per cent resident in towns of more than 5000 inhabitants; by 1911 these proportions had risen to 44.2 per cent and 38.4 per cent respectively (Pouthas 1956). Seine département had grown to house 4 154 000 people and

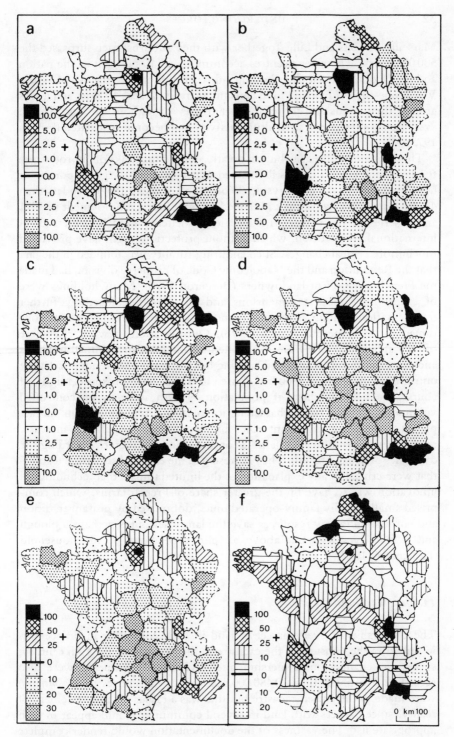

**Figure 3.1** Intensity of migration (expressed as a percentage of total population at start of period: (a) 1836–51; (b) 1851–66; (c) 1872–91; (d) 1891–1911; (e) 1836–1911. (f) Percentage population change, 1821–1911.

Marseilles, Lyons and Lille (together with their suburbs) each surpassed the 500 000 mark. These four centres accommodated 15 per cent of the nation and the next six largest cities housed a further million. On the eve of World War I, one-fifth of France's food consumers lived in these 10 cities and it was around their needs and those of lesser towns throughout the land that an ever-increasing share of agricultural activity had become articulated (Fohlen 1973).

This fundamental redistribution greatly swelled the ranks of unproductive food consumers but France still retained an exceptionally large proportion of her population in the countryside (Léonce de Lavergne 1857). No less than 44 per cent of the workforce remained in agricultural employment in 1911 and a series of defensive tariffs had long sheltered French farmers from international competition, which without protection might have provoked much more critical changes. Such enduring rurality was founded in the fact that the Revolution and the Napoleonic Code of property division had made the French a nation of landowners (Bergeron 1970). Many holdings were of exceedingly modest dimensions and all were threatened by further fragmentation so long as the inheritance laws were obeyed and heirs continued to be numerous (Ruttan 1978). Outmigration to work elsewhere offered an escape route from the limited resources of many small holdings and, in some instances, allowed farming heirs to lease land, if not purchase it outright, from relatives who had chosen to travel. Such prospects might, of course, be held in check if population growth were to be controlled voluntarily. These basic characteristics of peasant outmigration in fact played even more important roles by decelerating population increase, retaining a large number of families on the land, and thereby retarding the dynamics of urban growth. They represented the essential structural features that were crucial to any explanation of the limited impact that technological innovation was to have on the greater share of French farms, which comprised small, usually family-operated units, dominated by peasant tradition and commanding few resources save the land itself, plus perhaps a plough and a few livestock, but above all else the labour of the household (Hohenberg 1972).

## Holding the land

The ownership of every parcel of land in each commune in France was inscribed in the registers of the cadastre in the first half of the 19th century; successive changes in ownership and divisions of property were also recorded. No fewer than 10 480 000 landownership entries (*côtes foncières*) were made in the ancient cadastre but this was a rather inflated statement of reality, since owners with land in several communes would appear in each appropriate list. The vastness of the documentation would render complete cross tabulation of information between communes practically impossible for more than a local scale of evidence. The average entry from France as a whole was 4.6 ha, but it fell below 4.0 ha in départements across much of the more densely occupied northern third of the country, with holdings of less than 3.0 ha being characteristic of the Ile-de-France, Picardy, Artois and Flanders and reaching a minimum in Seine (0.65 ha) (Fig. 3.2a). Average entries for the plains of Caux and Champagne were close to the national mean, which was large in the context of northern France, and holdings were

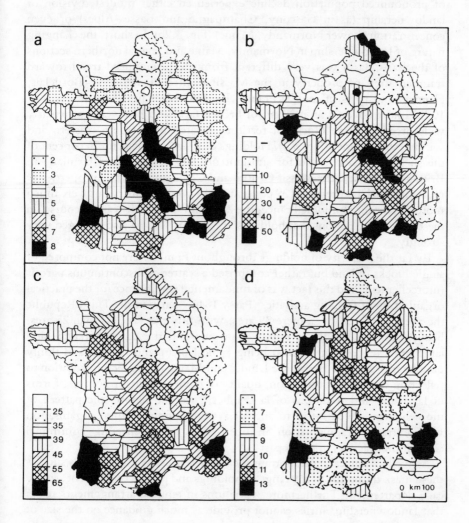

**Figure 3.2** (a) Average size of côtes foncières from ancien cadastre (ha); (b) change in number of côtes foncières, ancien cadastre–1907 (per cent); (c) *morcellement*, average size of parcels, 1882 (ares); (d) mean size of farm, 1892 (ha).

above average size throughout middle France and in parts of Armorica. Very large units were more typical of the Alps, the uplands of central France and the relatively empty lowland areas of Berry, Bourbonnais, Camargue and Landes (Devailly 1980).

By 1851 the total number of entries had reached 11 750 000 and rose to 13 050 000 in 1908, reflecting an increase of 24.5 per cent since the ancien cadastre (Burat 1851). The rate would undoubtedly have been greater had the tempo of population growth not slackened after mid-century. Indeed, areas of pronounced population decline experienced either modest division of landownership (as in Gascony, Champagne and Basses-Alpes) or even concentration (Lower Normandy, Maine) (Fig. 3.2b). In short, the changing nature of landownership in Normandy, Maine and several northern sections of the Paris Basin was very different from the pronounced trend toward fragmentation of property in the Massif Central, Languedoc, Bordelais, Flanders and parts of middle France. By 1882 côtes foncières of less than 10 ha apiece covered 35 per cent of French farmland but were particularly widespread in three blocks of territory located in northern, eastern and southwestern France (Fig. 3.3). Holdings of 10–40 ha made up 26 per cent of the surface but accounted for substantially larger sections of farmland in Brittany, Gascony, the Massif Central and Champagne; while properties of more than 40 ha represented 39 per cent of the whole country but were much more important in middle France, the Alps and Landes. It was, of course, in these very same regions that holdings in excess of 100 ha also emerged very strongly.

By far the majority of holdings throughout France were not composed of single blocks of land but rather comprised a scatter of discontinuous parcels (morcellement) and this fact was of fundamental significance for the practical organisation of farming activities (Passy 1846). Early in the Third Republic the average parcel was a mere 39 ares, with tiny parcels (less than 25 ares) being found in the Ile-de-France, Lorraine and the pays charentais and large units (exceeding 55 ares) being typical of upland regions and thinly populated areas, like Berry, the Landes and Maine (Fig. 3.2c). In addition to reflecting density of occupation, quality of environmental resources, forms of land use, and traditions of land inheritance and tenure, the pattern of morcellement owed much to ancient systems of agricultural organisation with extreme fragmentation in open field areas reflecting communal cultivation practices in times past.

As well as concealing important local differences, average figures such as these mask the nature of the range of holdings and parcels of different sizes in each département. Furthermore, variations in tenurial arrangements mean that landownership entries cannot provide as much guidance on the size of farm unit that may be expected. A number of cautious observations may be advanced, nonetheless. First, small units of landownership suggest pressure of population on land resources which might provide a stimulus for raising

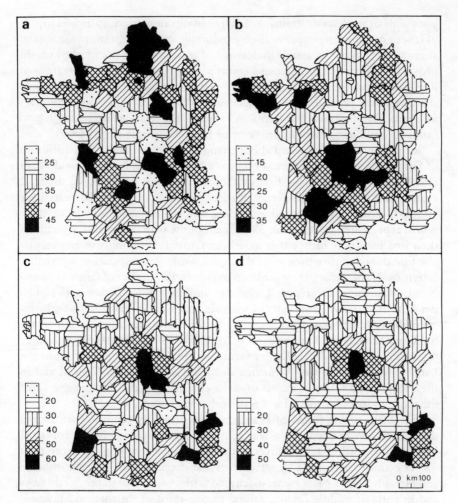

**Figure 3.3** Percentage of agricultural surface occupied by côtes foncières, 1882: (a) less than 10 ha; (b) 10–40 ha; (c) more than 40 ha; (d) more than 100 ha.

agricultural productivity to feed the farming family, to pay rents under conditions of tenancy, or, in less isolated areas, to orientate a part of production toward catering for urban demands, thereby generating cash that would be available for redirection into the local economy (Gobin 1859). Small units might well be suited to manual labour but they normally proved quite inappropriate for newer, mechanised technologies. Secondly, large holdings suggest lower pressure of population which reflect not only rather different tendencies of land occupation in earlier times but also perhaps differing combinations of environmental resources, possibly involving upland areas with harsh climates or difficult terrain, or lowland areas with poor soils or inadequate drainage. Such holdings might lack sufficient inputs of labour or capital to improve their productivity and might simply stagnate; alternatively, given accessible locations and adequate investment, they might prove appropriate for mechanised cereal production or specialised rearing of livestock on a large scale (Lebrun 1972). Thirdly, the operation of tenancy or sharecropping provides a set of mechanisms for subdividing large holdings into rather larger functional farms. As a result, the relationship between land holdings and farm units can prove to be far from simple (Zolla 1887).

Surprisingly little is known about changes in the number and distribution of farm units since only three surveys were undertaken before World War I (1862, 1882, 1892) and the first of these was not comprehensive. In 1862 France contained 2 950 000 farms of more than 1.0 ha apiece; 20 years later 5 450 000 farms were recorded, including miniscule units of less than 1.0 ha, which represented no less than 38 per cent of the national total. By virtue of their smallness they must either have been farmed exceedingly intensively or else functioned as part-time units that were worked in association with other activities. Nonetheless, they made up over half the number of farms in Seine, Nord, Meurthe and Pyrénées-Orientales and constituted between 45 and 50 per cent in Picardy, Languedoc, Lyonnais and middle France. It was, however, the farm of between one and 10 ha which formed the typical peasant holding of the Third Republic, accounting for no less than 46.3 per cent of the national total in 1882. Larger farms of 10–40 ha were important in the Landes, Seine-Inférieure, Armorica and parts of the Massif Central and in these same areas farms in excess of 40 ha comprised more than 5 per cent of the total, by comparison with 2.5 per cent nationwide.

Four-fifths of the farms of France were worked by families that owned them, and owner occupation was particularly widespread in the southern half and the northeastern quarter of the country, accounting for more than 90 per cent of farms in 16 départements and exceeding 97 per cent in Seine and Hérault (Fig. 3.4a). Owner-occupied farms were rarer in the north-west and in parts of middle France, falling to one-third of the total in Mayenne (33 per cent) and Seine-Inférieure (37 per cent). As the information on côtes foncières might lead one to suspect, the average owner-occupied farm was 4.5 ha in extent, with units of that kind of size being particularly characteristic of southeastern France and the eastern fringes of the Massif Central (Fig. 3.4b). By contrast with the very small units that typified parts of the Rhône valley, Aquitaine and Brittany, the average owner-occupied farm exceeded 6 ha in Champagne, Maine, parts of Normandy and much of the Ile-de-France, with very large units characterising poor environments in the Alps, Landes and Lozère. Intense morcellement was clearly not only the preserve of areas in which small farms predominated; indeed the small strips in the open fields of the Paris Basin and Lorraine were often the components of quite large family farms.

Tenancy (14.1 per cent) and sharecropping (6.4 per cent) together

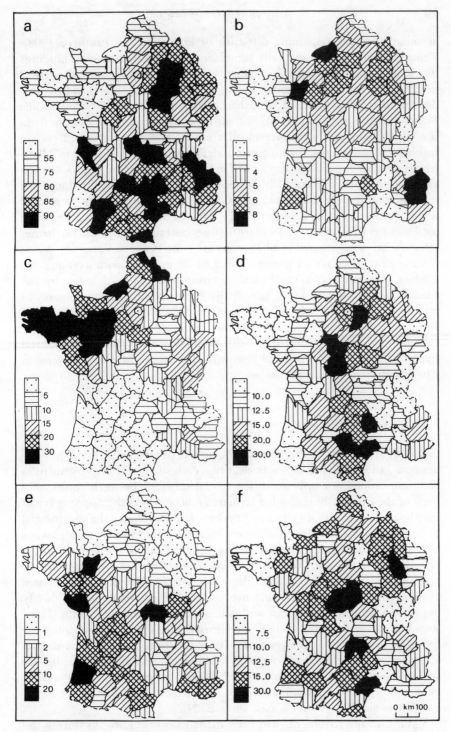

**Figure 3.4** (a) Owner-occupied holdings as percentage of the total, 1882; (b) average size (ha). (c) Tenanted holdings (per cent); (d) average size (ha). (e) Share-crop holdings (per cent); (f) average size (ha).

accounted for the remaining one-fifth of French farms (Prothero 1908). Tenanted farms were only really significant in the total number of holdings in a score of northwestern départements where the average size of tenanted unit was below the national mean of 12 ha (Fig. 3.4c). However, département means rose to double that size in parts of the Paris Basin, middle France and the south (Fig. 3.4d). Sharecrop farms were virtually absent from the Paris Basin and were rare in all areas save Bourbonnais and certain southern départements, forming just under one-quarter of all farms in Allier, Mayenne, and Vendée and almost half in Landes (46 per cent) (Fig. 3.4e). On average, they were fractionally larger (13 ha) than tenanted holdings and were markedly bigger in Berry and Champagne although they were very few in number in those provinces (Fig. 3.4f). In the subsequent 10 years the total number of farms rose by a mere 0.46 per cent to 5 481 695 and no further information was gathered until 10 years after the Peace of Versailles. Mean farm size, irrespective of tenure, was 8.65 ha in 1892, with averages exceeding 11 ha not only in the difficult environments of the Alps, Landes and Massif Central but also in the potentially more promising départements of middle France, Champagne, and Maine and the distinctly fertile pays of Beauce and Brie (Fig. 3.2d). At the other extreme, average farm sizes were exceedingly low in the immediate surroundings of Paris (Seine 1.97 ha) and Lyons (Rhône 4.32 ha) and fell below 7 ha in environments as diverse as Charentais, Dauphiné and Flanders.

## Degrees of accessibility

In the 1890s much of rural France was being served by the expanding railway network, and by that time it was clear that each successive improvement in communications not only modified spatial relations between component parts of the country but afforded farmers in distant areas the opportunity of supplying food to city dwellers and perhaps also challenged the commercial privileges of long-established tributary areas (Daudin 1834). New highways and canals had played their part during the first half of the century but it was the railways, with their unique speed and security, that worked in combination with feeder roads to allow the functional unification of French space gradually to occur. A truly national market emerged, at first slowly during the Second Empire but with added pace and certainty in the Third Republic (Toutain 1967). In 1840 a commission of enquiry, examining the proposal to build the railways from Tours to Bordeaux, had provided an accurate forecast of the shape of things to come. 'At present food supplies are sent to Paris from within a radius of 32–40 km. Open railway lines and the very same commodities will arrive from places 400 or even 480 km away, and just as quickly' (Wolkowitsch 1960, p. 27).

Significant improvements had been made to royal highways during the final decades of the ancien régime and the Paris Basin was endowed with a reasonably dense system as a result. But such was not the case across much of southern and central France, where highways were not only few but generally in poor repair. In the 1780s the fastest coaches had travelled between Paris and Toulouse or Marseilles in 8 days, by comparison with 12 to 15 days in the 1760s, but such speeds were only possible for affluent passengers; the movement of carts laden with grain or other bulky commodities was quite a different matter (Arbellot 1973). Minor tracks serving villages and farms were universally in poor condition, and years of neglect toward the end of the 18th century had made things worse. However, during the best years of Napoléon's regime important improvements were made. New roads and bridges were built, existing highways upgraded, canals constructed, and great quays laid out in Paris. Highways that were built for strategic purposes in Picardy and eastern France also served economic functions, as did new roads constructed at Napoléon's behest in the Ile-de-France, Burgundy, Dauphiné, Vendée and Brittany (Cavaillès 1946). The decree of 1811 established the critical distinction between national (or 'imperial') highways, to be built and maintained by central government, and secondary roads that would be the responsibility of départements through which they passed. These obligations were not met in the desperate months that followed and the legacy of the First Empire to subsequent administrations was one of neglect and degradation. In addition, the appalling roads of central France had been virtually ignored since so much of the energy of Napoléon's road builders had been directed to the Alpine passes, the Mediterranean coasts and the Pyrenees in order to provide links into Italy and Spain.

In 1828, 14 000 km of national ('royal') highway were still severely damaged, a further 500 km needed to be completed; and only 18 750 km were in good condition. The achievements of the Restoration proved to be modest but a more vigorous policy was pursued under Louis-Philippe, which was further enhanced by 1460 km of strategic roads built at the government's expense in eight western départements, following the uprising instigated by the Duchess of Berry in 1832. The 82 départements contained 32 980 km of national highway in 1837, which represented just under half the total network of main roads (Schnitzler 1846). Three-quarters of the royal highways were in good repair, by comparison with only two-fifths in 1824, but only 3100 km were paved and 20 500 km surfaced with stones. New legislation in 1837 initiated further investment to extend the national highways by 1500 km in 1855, when 98 per cent were deemed to be in good repair. Highways that were the responsibility of the départements totalled 36 100 km in 1837, but only two-fifths were in good repair and those in central France and southern France remained both rare and of poor quality. Everywhere, their impact on the surrounding countryside was slight, since most minor roads and tracks from villages and farms were in a deplorable state and generally proved impassable in wet weather (Bouchard 1834).

Legislation on local roads was introduced in 1824 which enabled communes to raise resources for *chemins dits vicinaux* but there was no compulsion to do so and few improvements were actually made. A more effective policy was facilitated by the law of 21 May 1836 which distinguished *chemins dits vicinaux de grande communication* and *routes départementales* from *chemins vicinaux de petite communication* (or *chemins communaux*). Communes and départements were to co-operate in constructing this final new category of road which was placed under the direction of the départements. Able-bodied taxpayers were required to contribute three days' labour each year or pay an appropriate tax for building and maintaining local roads. Prefects and conseils généraux responded quickly and in the same year a start was made on designating chemins vicinaux throughout France. Substantial sums were allocated for new roads and bridges following the law of 14 May 1837 and, as the practical implications of this legislation came to be felt, the French countryside began to escape gradually from centuries of isolation. But the enormous tasks of building and maintaining the local roads was to prove slow and difficult and it was not until the second half of the century that real progress was accomplished, with legislation in 1868 guaranteeing financial aid to communities that could not raise sufficient resources for essential road improvements (Maspétiol 1946).

Waterways provided the age-old means of transporting commodities in bulk during the ancien régime économique and 8500 km of rivers were deemed navigable in the 82 départements early in the July Monarchy, but conditions were poor along many stretches, with seasonally low water levels and the presence of shoals, mills and ancient bridges presenting serious hazards to boatmen. In addition, almost 1000 km of canal had been dug before 1789 and a further 200 km were excavated during the Napoleonic years. When the Allies evacuated French territory in 1818 no less than 3000 km of projected canals remained unfinished, but special loans were raised and building was accelerated, so that over 900 km of new canals were added by 1830. Construction continued apace and a total of 3500 km was recorded in 1837 but unfortunately many canals, new as well as old, were highly imperfect (Sée 1927). The Burgundy canal, which linked the Seine and Saône systems in 1833, was torrential in some periods of the year but virtually dry in others. The new Berry canal was usable for only two-thirds of the year because of ice or insufficient water, and the Rhône–Rhine canal, completed in 1832, was frequently short of water and contained numerous locks which hindered the flow of traffic. Many older canals were shallow and their banks were liable to collapse. Despite the important achievements of the July Monarchy, major gaps in the national waterway system remained: between the rivers Saône and Oise, lower Seine to lower Loire, between the Loire and Garonne, Bordeaux to Lyons, and from the Garonne to the Rhône (Chevalier 1838).

Some of these deficiencies were made good, albeit at a somewhat leisurely pace during the Second Empire and more vigorously during the Third Republic (Clapham 1921). As well as other formidable projects, Freycinet was responsible for an ambitious programme for improving navigation on 4000 km of river and 3650 km of canal and for building a further 1380 km of canal. These objectives were embodied in legislation of 1879 which stipulated that main waterways should conform to uniform standards (depth 2.0 m, length of locks 5.2 m, clearance of bridges 3.7 m). Further improvements were legislated for in 1903, and more work was accomplished on French waterways between 1879 and 1910 than in any other three decades, with 3300 km of channel being improved and 650 km of new canal being built. The volume of freight carried by waterway doubled over the same period. Building and maintenance of main roads continued apace after 1850, with the total length of national highways reaching 38 000 km in the 1890s, and increasing attention was also paid to extending and improving feeder roads. Important though these works unquestionably were, no really revolutionary change in speed or efficiency was achieved; that was only to prove possible with the construction of the railways.

The earliest lines had been built around Lyons and Saint-Etienne and in Languedoc during the late 1820s and 1830s (Anon. 1846). They covered short distances and were designed for carrying industrial raw materials and finished products, although they offered the potential for transporting cereals and wine to Saint-Etienne and into the Cévennes. The first line in the Ile-de-France linked the capital to Saint-Germain in 1837 but as early as 1830 Legrand, head of the highways department, had advocated a network of railway lines radiating from Paris to emulate the system of national highways. Parliamentary decisions in 1833 stipulated that concessions should be made by Chambers, not the Crown; money was duly set aside and the rudiments of a national network were proposed. Principal lines would radiate from the capital to the main ports (Le Havre, Calais, Bordeaux, Marseilles), neighbouring countries, industrial areas (Loire Basin, Nord, the north-east), and to the main fortresses (Metz, Verdun) and naval bases. In 1842 France contained only 570 km of track, by comparison with 3800 km in Great Britain, but legislation authorised a further 3600 km from Lyons to Mulhouse via Dijon, from Bordeaux to Marseilles via Toulouse, and between Paris and six provincial destinations. A veritable construction boom followed until the commercial crisis of 1847. By the end of 1843 Paris had been linked to Orléans and Rouen; four years later lines reached Le Havre and Lille, but the nation still possessed only 1900 km of track. Great debate surrounded the direction of future routes, and execution of the plan was not to come about until the Second Empire.

Napoléon III and his advisers favoured rapid construction, with major links from Paris to Strasbourg (1852) and to Belfort (1858) helping to raise the national total from 3550 km in 1851 to 8680 km in 1858, by which time

nearly 30 companies, which controlled about half the track, had been
brought together into six great regional consortia (Fig. 3.5). By the end of
the Second Empire the national network had doubled to 17 440 km and a
start was also being made on narrow-gauge 'railways of local interest'
(*tortillards*) to link villages and small towns (Caralp 1951). These had been
authorised as early as 1865 and were constructed by local authorities and
private companies. Laying wide-gauge tracks continued apace during the
Third Republic although it became clear that some existing lines and many
that were still only proposed would not be able to pay their way (Plessis
1973). Legislation of 1878 placed some lines under national administration
and made provision for the state to build a further 8850 km, thereby enabling
the Freycinet plan of 1858 for linking every sub-prefecture and prefecture to
the capital to be executed (Henderson 1967). By the eve of World War I the
total network had risen to 43 730 km and railways had been laid across
remote areas with difficult terrain, such as the Alps and southern Massif
Central, where new tracks were still being installed after 1900. In addition,

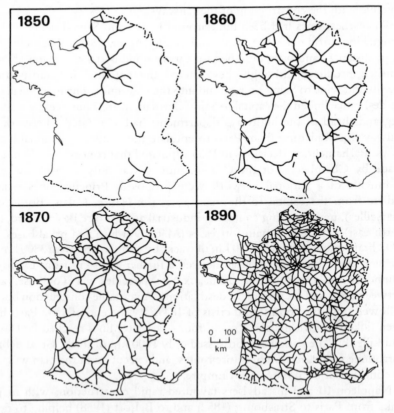

**Figure 3.5**  Development of the railway network.

the length of tortillards reached 1800 km in 1881 and increased to 9350 km on the eve of World War I, partly in response to local electoral pressures. Through careful planning a finely meshed, fairly evenly spread network of wide- and narrow-gauge lines had been established across the face of the country. Trains reached towns high in the mountains and villages deep in the heart of the country, but journeys from the provinces to the capital were always much simpler and speedier than comparable lengths of cross-country route. The centrality of Paris in the system was overwhelming.

## Widening horizons

Transport improvements in France were paralleled by the construction of railways in the New World and the operation of steamship routes to link Western Europe to the wider world. During the 1850s railway construction in North America allowed the wheatlands of the Great Plains to be opened up and in the following decades costs of bulk shipment by rail and steamer were cut substantially (Tracy 1964). The effects of the American Civil War (1861–4) curbed American wheat exports, but the peace that prevailed in America and Europe during the four decades that followed the Franco-Prussian War (1870–1) provided the necessary conditions for a great expansion in trade. North America benefited from four consecutive good harvests after 1877, but in Western Europe these were poor years with the harvest of 1879 proving particularly disastrous (Fig. 1.1). In the past, small harvests in Europe had stimulated higher prices which helped to maintain financial returns to farmers; now cheap American grain supplies could be brought in so that European prices fell rather than rose. From the 1870s onwards grain prices fell steadily in France since imports continued to flood in even after the recovery of domestic production in 1882–3 (Fig. 3.6). Indeed, the whole period from 1873 to 1896 was one of economic depression and consequent worsening of problems for French agriculture (Zolla 1899).

Exposure of French farming to overseas competition had been facilitated by the Anglo-French commercial treaty of 1860, whereby France had reduced tariffs on imports of manufactured goods and had practically erased protection for agriculture by removing import duties from nearly all foodstuffs and raw materials. Abolition of the sliding scale of duties on imported grains followed in 1861. This system of free trade came under severe strain after the Franco-Prussian War when a revival of nationalist sentiment ensued at the same time as a heavy burden of debt. A growing protectionist lobby was stimulated by industrial interests, which found an ally among the farmers who had suffered financially because of the poor harvest of 1879. Consequently a new tariff was adopted in 1881, which protected manufactures and raised duties on imported livestock; but cereal farmers remained vulnerable to cheap imports since wheat was subject to only a nominal duty

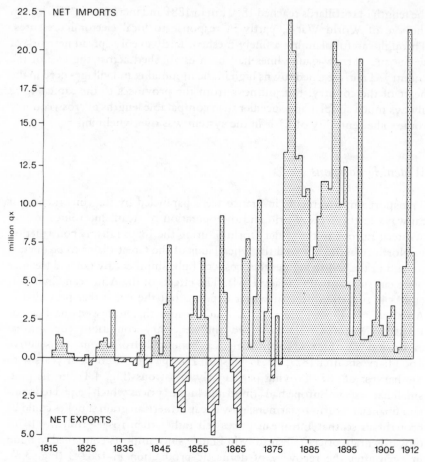

**Figure 3.6**  Wheat, net imports and exports, 1815–1912.

and other grains were exempt. Measures of protection were granted in 1885 and 1887 as duties on all major farm products were raised very substantially and grain imports were restrained. This permitted a recovery in French farm prices in the late 1880s, although they were to fall again in the early 1890s. The Méline tariff of 1892 reinforced the process of agricultural protection by increasing levies on many dutiable farm products (livestock, meat, cheese, wine, beer, hops) and placing duties on some goods that had entered free (maize, rice, potatoes, vegetables). Nonetheless, French farming remained plunged in depression because cereal imports continued to increase (Barral 1968). One response was to raise the duty on wheat once again and a further element of protection was provided by stopping cattle imports for health reasons between 1892 and 1903, whereafter a strongly prohibitive tariff was set in operation.

A gradual improvement of living standards throughout France and especially in her cities during the final quarter of the 19th century was reflected in an increasing demand for livestock products (Toutain 1971). Prices of meat and other livestock products were sustained higher and for a longer time than those of cereals, and as a result livestock producers and arable farmers were not affected equally by the effects of overseas competition. Livestock prices certainly did decline but to a lesser degree than those for cereals, and this meant that livestock farmers were able to benefit from substantial reductions in the price of feed grains. The net effect was to encourage a shift from grain production to livestock husbandry among at least some of the sections of the farming community that were gearing themselves to market demand. This trend was further reinforced by the prohibition of cattle imports after 1892. Indeed, France became self sufficient in cattle and beef after 1900, with occasional surpluses for export. The experience of French sheep rearers was just the reverse. Wool had been left free of duty in deference to the interests of textile manufacturers; imports increased substantially after mid-century, prices fell, and as areas of waste and fallow also retreated so the number of sheep in France declined. The fall would have been even greater had the consumption of sheep meat not increased.

During the 100 years from the end of the Napoleonic Wars to the outbreak of World War I France became more populous, urbanised and affluent. Roads, canals and especially railways served to break down the isolation of many areas that had been virtually ignorant of the food needs of the capital and other major cities in times past (Vigier 1963). At first, more land had been brought into cultivation with the first half of the 19th century, but this age-old response to rising population pressure was gradually replaced by a drive toward intensification and specialisation among a growing section of the farming population. Inward looking farms still remained extremely numerous in many parts of the country as late as the 1900s, but in some pays traditional forms of self sufficiency and mixed farming were being overtaken by enterprises that emphasised wheat growing, livestock rearing, viticulture and a host of other specialisms. The cumulative effect was to transform the face of much of the French countryside.

## Note

[1] The patterns shown on Figure 3.1 are derived from a reworking of census figures and of volumes of natural increase and net migration calculated by Pitié (1971).

# 4  Fluctuating fringes

## The old order changeth

In the early 19th century France still contained extensive stretches of land that remained uncultivated because of isolation, poor quality, steep slopes or communal status. But such areas of marsh, heather moor, stunted scrub or rough vegetation were far from being useless since they provided valuable pasture reserves and performed many other vital rolés. To many Bretons the moors were 'another Eden' and the furze that they produced was 'the veritable lucerne' of the peninsula and was 'as valuable as gold' (Tarot 1840, p. 540; Argouarch 1884, p. 7; Girard 1901, p. 44). Communal status performed the beneficial function of helping to conserve many fragile environments in areas such as Hautes-Pyrénées, but commonlands in lower, less vulnerable districts also displayed 'the sad spectacle' of extensive wastelands that stretched to the horizon around Lannemezan and Tarbes, thereby 'bringing dishonour' to the whole département, since many such districts were in fact suitable for cultivation (Inspecteurs: Hautes-Pyrénées 1843, p. 87). As De Gasparin (1846–7) explained, wastelands functioned as a 'safety anchor' for the traditional subsistence economy, forming a reserve of land that could be used in some form of extended rotation (as in the Champagne crayeuse, where rye and buckwheat were grown on temporary clearings) or else cropped on a rather more permanent basis during phases of rising demographic pressure (Hau 1976). Such was certainly also the case in northern Brittany and over parts of the 150 000 ha of Indre département that supported heather and other 'wild and useless plants' but also bore abandoned plough furrows (Inspecteurs: Côtes-du-Nord 1844, De Quincy 1831, p. 10). Unfortunately, paring and burning, followed by temporary cultivation, all too often gave rise to dire results on sloping stretches of land, as in the Montagne Noire where this 'savage and detestable system' was encouraging soil erosion and 'condemning agriculture to languish in its miserable *status quo*' (Inspecteurs: Aude 1847, p. 133). Many fertile pays also contained stretches of marsh or moor, as along the Scarpe valley in Artois where 10 000 ha were flooded every year and suffered very grave damage in the 'disastrous years' from 1826 to 1828 (Leroy 1829, p. 73).

In accordance with the spirit of the physiocrats, decrees and fiscal encouragements had been introduced to stimulate défrichement during the final decades of the ancien régime, but there is no way of knowing how much land was actually cleared since many declarations may have been statements of intent rather than fact (Sutton 1977). Nonetheless, it has been estimated that perhaps 300 000 ha might have been brought into cultivation between

1766 and 1789 (Castang 1967). Changes in landownership following the Revolution triggered off another burst of reclamation; however, application of legislation in 1792 to facilitate division of commonland was made optional in 1793 and progressively rescinded in 1795 and 1803 (Clapham 1921). As a result, communal property survived well into the 19th century in many parts of France (Sargent 1958). Clearing wasteland offered the chance of growing more basic food for human survival, which, of course, remained the essential objective of agriculture throughout much of France in the early 1800s. However, patches of commonland continued to be divided and sold prior to défrichement and Léonce de Lavergne (1861) insisted that such areas were 'without doubt richer and more populous' than those that retained their commons (p. 34). By contrast, in much of Brittany, Limousin and some other areas, the peasantry sought to defend the old order of land organisation and, in such localities, rural communities proved to be much more cohesive than in districts where commonland had been reclaimed (Corbin 1975). In short, wasteland survived for a long time in a large number of places, including the much researched commune of Plozévet in Lower Brittany, where the law of 1793 remained 'a dead letter' for a very long time, since a start on dividing the commons was not made until 1845 and real change occurred only after 1880 (Burguière 1977, p. 85).

In the first half of the 19th century, land clearance normally meant operating at a local scale, with an extra plot or two or a few new fields being added through the efforts of individual families or groups of farmers. In Armorica it often involved a modest co-operative effort, with a peasant who wished to clear land either inviting his neighbours directly or making use of the parish crier after Sunday mass (Anon. 1840–1). The act of défrichement was sometimes also a time of *fête*, with hard physical work being followed by food, drink and dancing to the sound of *biniou* pipes. Although the technology was simple and the scale remarkably local, the cumulative effect of such activities could be impressive, as irregular patchworks of fields won from the waste testify in the rural landscape in many parts of France. Of course, the whole business was frought with risk, since uncultivated soils were often inherently poor and certainly lacked the benefit of prolonged fertilisation which permanently cultivated plots had enjoyed.

In the absence of a rigorous enquiry prior to the ancien cadastre it is not possible to trace the progress of défrichement during the early 19th century; however, Lullin de Châteauvieux (1843) estimated that 200–300 000 ha had been reclaimed throughout France between 1789 and 1830, and numerous local instances of land clearance were recorded. For example, important progress was made in Perche during the 1790s, in the *touyas* of Basses-Pyrénées after 1810, and in the *brandes* of Poitou during the First Empire, with the encouragement of Prefect Cochon de Lapparent (Anon.: Orne 1802, Bottin 1833, Fénelon 1978). Likewise, many pools were drained in the Sologne between 1789 and 1830 and their sandy soil was devoted to rye and

buckwheat since it proved unsuitable for wheat, at least in the first instance
(Doé 1832). Unfortunately, some of these areas were so infertile that they
had to be abandoned for 10 years or more after just one miserable crop had
been harvested (Lefebvre 1962). Communal marshland was being divided
along the Atlantic coast and in the Camargue, with the help of special
syndicates which organised the task of reclamation (Sermet 1930, George &
Hughes 1933). In many instances défrichement had profound implications
on the rural economy at large as in the Corbières (Aude) where the number
of fine-fleeced sheep fell from over 1 million in the 1780s to under 600 000
in 1812 because sections of *garrigue* had been reclaimed (Trouvé 1819).
Division and clearance of common grazing land produced even more dire
results in Ariège, where serious soil erosion ensued and the uncultivated
patches which remained were unable to provide sufficient fodder for
satisfactory grazing to take place (Mercadier 1802).

In addition, the 19th century witnessed another scale of défrichement
which involved ambitious schemes and wide expanses of land and therefore
demanded more than local initiative. Much greater support was required for
financing new forms of communication, guaranteeing necessary supplies of
fertiliser, and ensuring adequate water control. In short, a completely new
technology was needed to tame the extensive wastelands of Armorica,
middle France, Aquitaine and the Massif Central. During the ancien régime
the physiocrats had outlined the agricultural potential of such areas and some
large landowners duly rose to the challenge in the first half of the 19th
century. In particular, the émigration intérieure at the start of the July
Monarchy brought many legitimist landowners back to their country estates
but it was not until after mid-century that a series of circumstances con-
verged and encouraged large reclamation schemes. Thus, construction of the
railway network reflected the emerging unity of France and accentuated the
political requirement for administrations to be seen to be taking notice of
economic and social problems in all parts of the land (Girard 1952). Heavy
investment in the railways and the modernisation of Paris made it all the
more imperative for Napoléon III to ensure that funds were also devoted to
the countryside.

In 1852 the central body of the Ponts-et-Chausées drew attention to the
pitiable state of cultivation, the extensive wastes and lack of capital for land
improvement in the Sologne, Dombes, Landes and other mauvais pays. A
programme of interior colonisation was promptly drawn up, with direct investment from the central administration. In addition, the *crédit foncier* was
created to help finance other rural improvement schemes but it was to prove
poorly suited to the needs of owners of small and medium sized properties
(Goujon 1976). Likewise, the government loans that were launched in 1856
to encourage piped underdraining were to attract few takers. Important
amounts of défrichement were accomplished during the Second Empire
nonetheless, but such achievements were soon to be overtaken by a very

different combination of circumstances whereby farmland started to be abandoned in response to agricultural depression or severe depopulation. As a result, the national total of uncultivated land began to increase once again.

The diverse and somewhat ambiguous nature of 'wasteland' renders it not only a difficult form of land use to conceptualise, but also one that was defined rather differently in the ancien cadastre and the enquête agricole of the July Monarchy. This fact produced substantial variations in the amounts cited in these roughly contemporaneous sources (Tables 4.1 & 2). According to the ancien cadastre, France contained 7 588 735 ha of wasteland and that total was to decline to 7 185 875 ha in 1851 and 6 230 500 ha in 1879, and rise thereafter to 6 446 810 ha in 1907. In fact, the ancien cadastre figure was 11 per cent below that in the enquête agricole of the late 1830s; the cadastral revisions were 14 per cent and 9 per cent respectively above those in the agricultural returns of 1852 and 1882; while the 1907 cadastral revision was 1.3 per cent below the combined agricultural statistics for *landes*, *terres incultes*, *pâturages* and *pacages* in 1912. Bewildering deviations between the two series also emerge at the département scale; however, the broad trend was clear enough. Scrutiny of data in the cadastral series suggests that most extensive défrichement occurred between 1851 and 1879, when the wastelands of France were retreating by an average of 34 000 ha each year. Annual net losses during the 1820s and 1840s had amounted to less than 20 000 ha, while after 1879 the uncultivated total was to increase by 7500 ha each year.

**Table 4.1** National land use from cadastral data (ha).

|  | Ancien cadastre | 1851 | 1879 | 1907 |
|---|---|---|---|---|
| wasteland | 7 588 735 | 7 185 875 | 6 230 500 | 6 446 810 |
| arable | 24 636 900 | 24 987 750 | 25 361 565 | 23 206 960 |
| grass | 4 612 455 | 4 591 430 | 4 804 780 | 6 664 280 |
| vines | 2 073 870 | 2 142 810 | 2 282 300 | 1 479 035 |
| woodland | 7 673 555 | 7 652 645 | 7 926 000 | 8 216 135 |
| all woodland★ | 8 572 850 | 8 500 630 | 8 878 375 | 9 257 580 |

★Ancien cadastre/1837;   1851/1862;   1879/1882;   1907.

**Table 4.2** National land use from agricultural data (ha).

|  | 1837 | 1852 | 1862 | 1882 | 1892 | 1902 | 1912 |
|---|---|---|---|---|---|---|---|
| wasteland | 8 563 480 | 6 283 785 | 5 976 875 | 5 721 685 | 5 706 060 | n.d. | 6 529 600 |
| arable | 24 523 910 | 25 341 450 | 25 590 945 | 25 366 555 | 25 140 165 | n.d. | 23 194 840 |
| grass | 4 044 645 | 4 892 940 | 4 788 715 | 5 529 095 | 5 922 010 | 6 654 895 | 6 232 100 |
| vines | 1 937 135 | 2 148 180 | 2 243 285 | 2 143 235 | 1 752 120 | 1 752 120 | 1 590 695 |
| woodland | 8 272 070★ | n.d. | 8 585 340 | 8 927 675 | 8 993 987 | n.d. | 9 333 710 |

★Including 335 730 ha of *sol forestier*.

Enduring land hunger during the July Monarchy promoted further attack on uncultivated spaces and the return of landowners imbued with 'improving' ideas provided additional stimulus. Small patches were reclaimed throughout the country and more impressive achievements were accomplished in Vendée and other pays on the fringe of Armorica (Le Boyer 1832, Labande 1976). For example, the brandes of Poitou retreated substantially during the 1830s (with use of lime and marl being introduced to improve the soil) and the foundation of an agricultural college at Grandjouan by Jules Rieffel in 1830 created a nucleus of wasteland clearance, land drainage and afforestation in Loire-Inférieure (Léonce de Lavergne 1861, De Penanster 1887). Rieffel and his pupils provided powerful examples of the practice of improvement, introduced Dombasle ploughs to work newly cleared soil and made use of *noir animal* from sugar works in Nantes to help fertilise it. Indeed, by mid-century noir animal was to become used quite widely in Allier and Indre-et-Loire as well as six départements around the lower Loire (Puvis 1850). Unfortunately, reclamation was sometimes undertaken without adequate appreciation of the local environment, as for example in Lot where extremely poor land was cleared which soon proved quite unsuitable for sustained cropping (Laffargue 1839).

During the July Monarchy the wastelands of France retreated by an average of 19 185 ha each year, although there were important deviations from this trend, with a dozen départements actually recording net increases in their landes. Admittedly, six of these districts involved less than 500 ha apiece, but sizeable increases were noted in Morbihan (up 6065 ha), Landes (up 8200 ha) and Ariège (up 16 140 ha). It is hard to envisage that farmland actually reverted to waste during this phase of vigorous population growth in the countryside and it is possible that this anomalous trend may have been due to reclassification of land by cadastral surveyors. Thus it is conceivable that degraded woodland or very poor permanent pasture may have been redefined as 'wasteland' in some départements at mid-century. Changes from one land-use category to another may not be traced explicitly; all that can be done is to identify which elements underwent substantial modification and then infer the nature of the relationship between them. Thus, while wasteland was increasing between the ancien cadastre and 1851, woodland underwent substantial decline in both Ariège (down 12 640 ha) and Landes (down 3210 ha), while in Morbihan permanent grass contracted by 8210 ha.

Taking data at face value, wasteland declined in no less than 70 départements during the 1830s and 1840s; and when miniscule increases are discounted the weight of evidence in favour of défrichement becomes virtually unanimous. In 13 départements more than one-fifth of the wasteland recorded in the ancien cadastre had disappeared by 1851, and in a further 15 départements between 10 and 20 per cent had gone. In 19 départements of western and southern France uncultivated land was retreating by 250–500 ha each year, reaching 500–1000 ha in seven more, and exceeding 1000 ha p.a. in Vienne (down 1125 ha p.a.), Vendée (down 1325 ha p.a.), Loire-Inférieure (down 1680 ha p.a.) and Hautes-Alpes (down 4030 ha p.a.) (Fig. 4.1a). Arable was the leading land-use category to increase at this time in eastern Armorica and parts of middle France, while vineyards represented the most extensive category to advance in départements along the Mediterranean littoral.

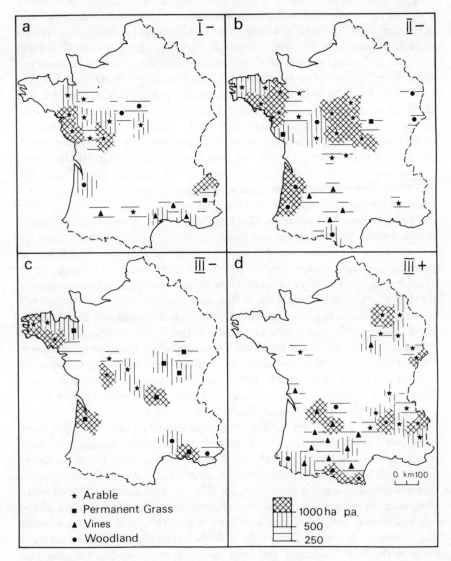

**Figure 4.1**   Wasteland, annual net change by cadastral phase.

## Interior colonisation

The Second Empire encompassed a phase of intérior colonisation that was stimulated, among other factors, by rising cereal prices, continuing vulnerability of food supplies (as the shortage of 1846–7 had so painfully shown), and government concern that France was lagging behind her neighbours in terms of agricultural development. Following the example of Belgium and Great Britain, the new French administration launched a series of legal, financial and practical measures to encourage drainage and défrichement (Clout 1977a, Dumas 1851). A hydraulics service was created in each département in 1848, under the direction of a Ponts-et-Chaussées engineer with the express task of investigating uncultivated land and making recommendations for reclamation. To transform the 130 000 ha of heaths and 13 000 ha of lakes in the Sologne formed a prime objective, which had attracted attention over several decades and especially since the émigration intérieure of 1830. According to Léonce de Lavergne (1861) the 'uncultivated steppes' of the Sologne were so disgraceful that any visitor might be excused for thinking he had returned to rural life in the Middle Ages. Numerous improvements had been proposed during the July Monarchy, including afforestation of very poor soils and construction of canals and railways for transporting marl and fertiliser prior to bringing areas of better soil into permanent cultivation. Work had started on an agricultural canal in 1848 but money soon ran out and there was underlying opposition to such a scheme since the canal, cut through the heart of Brittany, had failed to produce any rapid reclamation of the peninsula's wasteland. On the other hand, the idea of a railway to serve the Sologne rapidly gained in popularity. In 1859 the Comité Central de la Sologne was established with imperial backing to encourage land improvement, with drainage and afforestation on the Emperor's own estate at La Motte-Beuvron providing a practical example, which was soon emulated by surrounding landowners (Dupeux 1962, Marcilhacy 1962). In order to facilitate reclamation a network of agricultural roads to serve marl pits, railway stations and navigable waterways was proposed by the Ponts-et-Chaussées two years later and was eventually implemented (Sutton 1969, 1971, 1973).

Further to the west, the Brenne comprised 21 450 ha of heath, 6580 ha of lakes and 15 000 ha of woodland, as well as patches of uncultivated land which covered a further 60 000 ha (De La Véronne 1971). As in other marshlands fevers were rampant and death rates were higher than in surrounding districts (Le Bon 1862). In 1851 the hydraulics service of Indre instigated an inexpensive plan for clearing out watercourses, establishing reservoirs for irrigation and building agricultural roads to assist défrichement. In eastern France the reclamation of the Dombes had started early in the July Monarchy and by mid-century draining and liming had begun to make progress. Wheat had replaced rye in some localities, and clover had been introduced, which allowed more livestock to be kept and more manure to be generated. But serious problems would remain so long as the lakes of the Dombes continued to be used in their traditional way, being water-filled for two or three years and then drained and cropped with oats (Pacoud 1805). The dampness of the pays provoked disease and high mortality, supplies of lime and fertiliser were not available locally, and communications were inadequate to allow them to be transported cheaply (Guyétan 1856). In order to master these problems a full network of agricultural roads across the 100 000 ha of the Dombes was advocated in 1856. The scheme ran into difficulty since local communes refused to make land available free of charge, and landowners feared that défrichement would involve them in great expenditure. However, six years later the administration of Ain was allocated finance to build 240 km of agricultural roads (which were duly completed in 1868) and a concession was granted for a railway between Sathonay and Bourg which would greatly assist the transportation of fertilisers (Fléchet 1967). No less than 9000 ha of the 19 000 ha of lakes were brought into permanent cultivation during the 1860s, permitting a great improvement in the health of the local people (Marot 1958).

In southwestern France the vast stretches of the Landes contained 400 000 ha of dunes, pools, poor woodland and generally unproductive territory. Attemps to reclaim land and improve farming and forestry had been made earlier in the century, but with little effect since local residents fiercely defended their traditional ways. By 1855 only 28 000 ha of communal wasteland had been sold but, in that year, the Bordeaux–Bayonne railway was constructed through the area and a start was also made on a network of agricultural roads (Cavaillès 1933). Two years later the state passed special legislation to encourage land draining, road building and afforestation, with the Emperor once again providing an example on his estate of Solférino (Heuzé 1868). Municipalities were authorised to sell part of their common-land to help cover drainage costs and were entitled to seek free advice from state officials on proposed engineering works. Many local schemes were duly completed and by 1865 a canal had been dug behind the coastal dunes to void away surplus water into the Gironde or the basin of Arcachon.

At the end of the July Monarchy there had been only one stone-surfaced road in the whole pays and it normally took 15 hours to travel the 50 km between Bordeaux and Arcachon along a track of shifting sand. Conditions improved rapidly following the law of 1857, with the state building 22 agricultural roads in the next three years and an important start also being made on main roads. Medical officers dispatched encouraging reports on improved health conditions and substantial increases in life expectancy. For example, in 1865 the local doctor at Audenge reported that 'marsh fevers are now no more common in the Landes than in healthier areas of France. Until

1857 I used to administer about 1 kg of sulphate of quinine to my patients each year. Five years ago I bought 500 g which I still have in my possession' (cited in Huffell 1904, p. 205). Similarly, Dr Trelat could report at the end of a special investigation in 1878, 'there is no more pellagra in the Landes and the fevers have disappeared; the population now lives in clean and healthy villages, in pleasant houses surrounded by luxuriant vegetation' (cited in Chambrelent 1887, p. 65). In identical ways, the much smaller wasteland *pays* of the Double was endowed with a network of agricultural roads during the Second Empire and drainage of this part of the middle Garonne Valley was undertaken successfully from a base at the abbey of Echourgnac (Fayolle 1977, Livet 1942).

Each of these improvement schemes involved substantial conversion of waste to other forms of land use, most notably ploughland and forest, and together they made a critical contribution to refashioning the rural landscapes of many districts of France in the middle decades of the 19th century. Unlike the Sologne or the Landes, no grand design was drawn up for the extensive moors and heaths of Brittany but legislation was passed in December 1850 to encourage division, sale and reclamation of wasteland in the peninsula (Guellec 1979). Some 692 commons, covering 36 000 ha of wasteland, were subsequently divided and important défrichement was started in many localities (Chombart de Lauwe 1946, Clout 1973–4). Further south, the complicated and devastating impact of phylloxera was being experienced during the 1860s, bringing ruin to many vineyards and provoking substantial transformation of land use in its wake. Net increases in wasteland were indeed recorded between 1851 and 1879 in 12 départements, which were predominantly in southern France. The creative conquest of land was reckoned to be over as early as 1860 in much of Provence, with terraces and fields starting to revert to rough grazing or scrubby woodland as depopulation began to set in (Livet 1978). In the three decades following midcentury particularly large increases in wasteland were noted in Gard (up 30 650 ha) and Hérault (up 5490 ha), where viticulture had expanded massively earlier in the century only to be succeeded by devastation and land abandonment after the onset of phylloxera in the mid-1860s.

Between 1851 and 1879 wasteland contracted by more than one-fifth in 30 départements and by 10–20 per cent in a further 17, with the national rate of decline having almost doubled after mid-century to reach 34 120 ha p.a. The most extensive reductions affected middle France, eastern Brittany and western Aquitaine, with annual reductions of more than 250 ha occurring in 28 départements and peak values being recorded in Loire-Inférieure (3055 ha p.a.), Indre (2615 ha p.a.), Gironde (4900 ha p.a.) and Landes (8690 ha p.a.) (Fig. 4.1b). By comparison with the July Monarchy the foci of wasteland conversion in western and middle France had expanded substantially, but that in Languedoc had disappeared completely (Bobin 1926). Arable appears to have been the main beneficiary from défrichement in

virtually all départements of Armorica and middle France, although woodland was the leading category to advance in Aquitaine.

According to the agricultural enquiry of 1862, 229 865 ha of wasteland had
been cleared during the preceding 10 years, with peak values involving
Gironde (15 320 ha), Loire-Inférieure (14 485 ha), Indre (13 095 ha) and Ille-
et-Vilaine (12 135 ha) but some défrichement occurred in every département
in the land. Over France as a whole, 3.2 per cent of the wasteland recorded in
the cadastral revision of 1851 had been subject to défrichement in the following 10 years, with 13 départements in middle France (from Loire-Inférieure
to Nièvre and from Orne to Indre) having more than 10 per cent of their
waste reclaimed. By contrast, the heaths and moors of the Massif Central
and southern France had declined much less. Drainage of 57 035 ha of
marshland further contributed to the retreat of uncultivated land between
1852 and 1862, with seven départements in both coastal and interior locations
recording drainage of more than 1000 ha apiece and most being achieved in
Nord (3375 ha), Landes (4455 ha) and Gironde (30 345 ha) (De Buffon 1852).
Land drainage alongside the Gironde proved somewhat unusual since part of
the reclaimed surface was destined to grow vines, but it was like all other
schemes by virtue of the new employment that it created (Dupont 1847).

Intermittent fever and shortage of fodder during the summer were
problems that afflicted many marshlands along the west coast, and much of
the drainage accomplished there during the 1850s simply represented the
most recent attempts in a long saga of effort to cope with these and other
problems (Clout 1977b). No fewer than 120 syndicates of landowners
existed around La Rochelle to manage 72 000 ha of damp, low-lying land and
further north in Vendée 49 syndicates controlled 100 000 ha of marsh
(Labande 1976). Division of communal marshland and subsequent absentee
landlordism had hampered the work of these associations but such problems
were minimised by a number of rulings in 1824, 1865 and later dates which
enabled a majority of landowners to take the decision that a stretch of marsh
should be reclaimed (Bouscasse 1867). Important concessions along the bay
of Mont-Saint-Michel had been made to the Société des Polders de l'Ouest in
1856 and in the following 20 years over 1000 ha were reclaimed for cultivation (Rousselle 1877). Work continued in the interior valleys of Flanders
and the *wateringues* of the coastal plain, while in the Midi, drainage engineers
were confronting the additional problem of salinisation in the Camargue
(De Gasparin 1851). Unfortunately, the absence of agricultural enquiries
makes it impossible to determine the detailed progress of either défrichement
or drainage between 1862 and 1882.

## Full circle

The final decades of the 19th century and the early 1900s saw a major
reversal in the national trend of land-use change, with France containing

significantly more uncultivated land in 1907 than in 1879. Yet the process of défrichement had certainly operated in every département in the land between 1882 and 1892, with a total of 127 360 ha being put freshly into cultivation. Unlike the large clearances of the 1850s, only one département (Morbihan, 10 115 ha) experienced really extensive défrichement but 5000–10 000 ha were reclaimed in six others located in Brittany, Limousin and the pays of the middle Loire (Clout 1979). Just 2 per cent of the total wasteland recorded in the cadastral revision of 1879 was removed between 1882 and 1892 but in eight départements, including six in middle France, more than 10 per cent of the wasteland surface disappeared. In only 12 départements was more waste cleared during the 1880s than in the 1850s, notably in Brittany, northern Limousin, and a scatter of locations in Champagne, the Massif Central, Pyrenees and the Mediterranean littoral.

The effects of legislation in 1850 and 1860 for dividing and reclaiming commonland were experienced in various parts of France throughout the Third Republic as communications were being improved substantially to allow lime and fertilisers to be transported into previously isolated districts. For example, a number of large landowners purchased extensive areas of former commonland in Limousin during the early 1880s, organised their clearance (often with the help of teams of Auvergnat *défricheurs*), and undertook liming and fertilising prior to setting out arable fields or improved grass (Perpillou 1940). Elsewhere in central and western France the hinge years between the two centuries formed an important phase of reclamation which was quite at variance with the general transformation of land use in most other regions (Corbin 1975). In addition, 9755 ha of marshland were declared to be drained during the 1880s (much less than in the 1850s), with major achievements in Gironde (700 ha) and Loir-et-Cher (635 ha). In spite of all this work, many of the management problems that had long typified undrained marsh were still to be found in the final decades of the 19th century. Large stretches of excessively humid land remained along many coasts and estuaries, and the northward flowing rivers of inner Flanders continued to flood each year. Drainage syndicates in Nord had become apathetic, allowing watercourses to become clogged with mud, and it was clear to some that diplomatic contacts with Belgium needed to be enhanced since international action was necessary in this area (Anon. 1894).

Despite many recorded examples of land clearance between 1879 and 1907, the national trend was for more farmland to be abandoned than was being reclaimed, and as a result uncultivated land increased by an average of 7460 ha each year. Nonetheless, 29 départements contained less waste in 1907 than they had three decades earlier and in four of these areas over half of the 1879 total of uncultivated land had been converted to other uses. In 16 départements more than 250 ha of wasteland disappeared annually, with over 2000 ha going each year in Puy-de-Dôme (2450 ha) Bouches-du-Rhône (2120 ha), Finistère (2450 ha) and Morbihan (2785 ha) (Fig. 4.1c).

Défrichement made major advances in Lower Brittany where arable land was the leading category to increase, although that role was taken by permanent grass in Auvergne and Burgundy, and by woodland in the north-east. The remaining départements approached World War I with more uncultivated land than in the early years of the Third Republic. In eight départements the wasteland surface increased annually by more than 1000 ha, with a very great deal of farmland being abandoned in Lot (2380 ha), Ariège (2695 ha) and Hautes-Alpes (2920 ha) (Mergoil 1978) (Fig. 4.1d). As a result, two main regions of abandonment emerged; the first covering much of the southern third of France, and the second involving Champagne and Lorraine. Such increases in wasteland accrued partially from rural depopulation, as in Aquitaine and Upper Provence, but also from the decline of earlier viable forms of agricultural activity (Chombart de Lauwe 1946, De Réparaz 1966). The essential clues for elucidating this situation may be derived from the major land-use elements that underwent decline at this time. Thus, ploughland was retreating markedly in northeastern France, the Alps and the southern Massif Central; while woodland formed the leading feature to decline in the Pyrenees; and that role was filled by the vine in Gascony, Dordogne and Yonne following the ravages of phylloxera (Higounet 1978).

From Waterloo to World War I the marshes, moors and heaths of France clearly experienced strikingly varied fortunes. New fields and farmsteads, stretches of *bocage*, terraced hillsides, higher limits of cropping, reclaimed marshland and many other features formed tangible expressions of défrichement along an advancing fringe of cultivation. But the final decades of that broad span of years saw the appearance of numerous abandoned landscapes, even though protective commercial legislation had undoubtedly served to cushion many rural areas from even more dramatic decline. Thirty-six départements, mainly in the south and north-east, contained more uncultivated land in 1907 than at the time of the ancien cadastre (Fig. 4.2a). In Ariège, Hautes-Alpes and Lot the uncultivated surface had increased by more than 50 000 ha and by 25–50 000 ha in Ardèche, Dordogne, Drôme, Marne and Pyrénées-Orientales. On the other side of the equation, waste-land was less extensive in much of northern, northwestern and middle France, with important outliers of clearance in the lower Rhône and Aquitaine. The volume of decline involved less than 10 000 ha per département across much of the Paris Basin and Normandy and on the fringes of Armorica, but reached massive proportions in Brittany, middle France and Aquitaine, where net reductions in wasteland of more than 50 000 ha were encountered in many départements and absolute maxima were recorded in Loire-Inférieure (105 280 ha), Morbihan (105 600 ha), Gironde (191 270 ha) and Landes (238 580 ha).

This complex evolution may be summarised by a simple chronological typology which identifies the direction of change during each cadastral phase

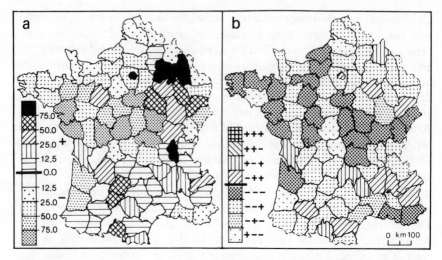

**Figure 4.2** Wasteland: (a) net change in surface, ancien cadastre–1907 (per cent); (b) typology of change.

but ignores both the volume and the rate (Fig. 4.2b). The national trend for wasteland to retreat until 1879 and increase thereafter (−−+) was encountered in 38 départements, almost half of the country (Table 4.3). Nine, predominantly southern, départements diverged from this trend and displayed reversion to waste from mid-century onwards (−++); intense rural depopulation or the collapse of viticulture or some other major activity may have triggered this precocious trend. Similar circumstances during the Second Empire may well have worked to interrupt an otherwise consistent trend of défrichement (−+−) in three other départements. Land clearance exceeded reversion during each cadastral phase (−−−) in 22 départements, with much of middle France, Armorica, Gironde and parts of Provence conforming to this pattern. Five other départements experienced an initial phase of abandonment before embarking on a prolonged period of clearance (+−−) and were so distributed as to fill out the regions of défrichement in western and middle France. The precise uses to which former moors and

**Table 4.3** Typology of national land-use evolution, from cadastral data.

|              | +++ | ++− | +−+ | −++ | −−− | −−+ | −+− | +−− |
|--------------|-----|-----|-----|-----|-----|-----|-----|-----|
| wasteland    | 0   | 0   | 5   | 9   | 22  | 38  | 3   | 5   |
| arable       | 13  | 32  | 4   | 3   | 7   | 2   | 1   | 20  |
| grass        | 31  | 3   | 11  | 18  | 1   | 15  | 1   | 2   |
| vines*       | 0   | 26  | 4   | 0   | 17  | 1   | 14  | 9   |
| all woodland | 15  | 5   | 8   | 13  | 12  | 16  | 10  | 3   |

*Not present in every département.

heaths were subsequently put may not be traced from cadastral statistics but may be surmised with the help of other contemporary evidence. Afforestation was the fate of former wastelands in the Landes and other mauvais pays, while large sections of the south were planted with vines (for a time at least), but by far the greatest share of land must have been used to grow more grain, since this remained the major imperative for the majority of family farmers throughout the 19th century as it had been since time began (Pautard 1965).

# 5 *Pivot of the economy*

## Advance

The resources of the arable realm unquestionably formed the key to human survival during the ancien régime économique. A wide range of cereal crops was produced in response to differing environmental and commercial conditions, and a sizeable total of livestock was also supported by grazing stubble and fallow land (Mulliez 1979). Application of animal manure was essential if the soil was to maintain its productivity at this time, when artificial fertilisers were largely unknown. These functional imperatives gradually changed as communications improved during the course of the 19th century and many farmers were released from the need to produce grain, becoming free to concentrate on other forms of agricultural activity. This complex transition involved important adjustments within the arable domain and between ploughland and the other land-use realms. Hence, ploughland retreated substantially in many parts of France after mid-century as animal husbandry, viticulture and other forms of production became more profitable than growing grain. Elsewhere population pressure declined markedly and, as a result, stretches of cultivated land reverted to waste or were planted with timber. In addition, technological advance enabled farmland to be rendered more productive, as soil moisture was controlled, fertilisers were applied, and mechanisation was developed to complement human labour, on the most progressive farms at least.

Despite centuries of human occupation and the operation of vigorous défrichement since the 1760s, it was possible for the arable surface of France to increase by over 700 000 ha from the time of the ancien cadastre (24 636 900 ha) to 1879 (25 361 565 ha). Thereafter the amount of land under the plough contracted dramatically to 23 206 960 ha in 1907. The nation's arable realm declined by 5.8 per cent over the full span of years but there were, of course, important spatial variations in both the direction and chronology of change. Thus, 22 départements in western, southwestern and middle France contained more ploughland in 1907 than at the time of the ancien cadastre, with arable increasing by 36 per cent in Loire-Inférieure and by more than 20 per cent in five other départements (Fig. 5.1a). In absolute terms Loire-Inférieure enhanced its arable surface by 112 310 ha with the neighbouring départements of Vienne (up 88 055 ha) and Indre (up 87 450 ha) also undergoing very important net increases. However, the experience of Armorica, middle France and Gascony was quite exceptional since arable land contracted throughout the remaining three-quarters of the country.

In nine départements ploughland declined by over 30 per cent, with

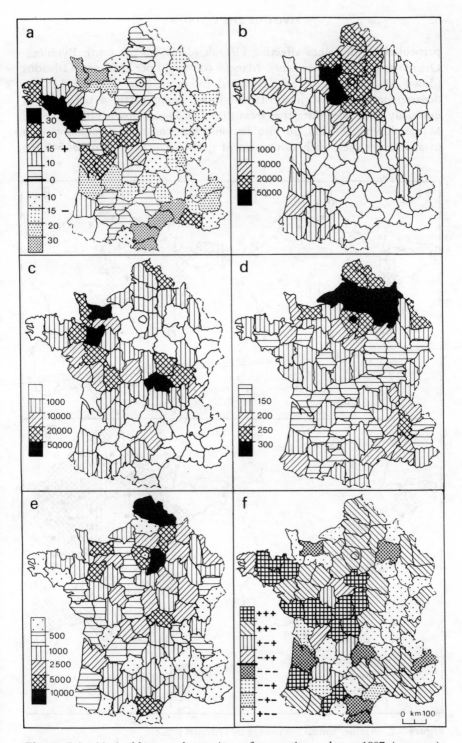

**Figure 5.1** (a) Arable, net change in surface, ancien cadastre–1907 (per cent); (b) land marled, 1852–62 (ha); (c) land limed, 1852–62 (ha); (d) application of animal manure, 1862 (qx/ha); (e) land underdrained during 1852–62 and 1882–92 (ha); (f) arable, typology of change.

particularly high values affecting Hérault (down 60 per cent), Pyrénées-Orientales (down 56 per cent), Manche (down 40 per cent) and Calvados (down 38 per cent). Arable retreated by more than 75 000 ha in six départements, with such substantial losses being partly a reflection of unusually large administrative areas (like Dordogne, down 110 720 ha; or Marne, down 98 245 ha), but also demonstrating the intensity of conversion to alternative uses, such as permanent grass (Manche, down 154 540 ha;

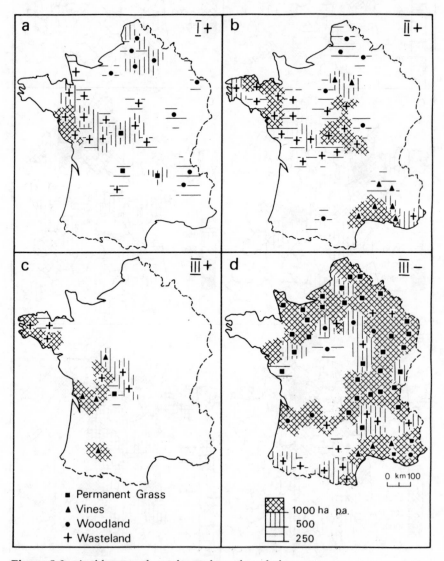

**Figure 5.2**  Arable, annual net change by cadastral phase.

Calvados, down 122 605 ha) or viticulture (Hérault, down 94 300 ha; Aude, down 92 265 ha). Sharp contrasts existed between these six départements and the districts that surrounded them, where only small reductions or even net increases in the arable surface occurred. For example, the pronounced retreat of ploughland in Lower Normandy contrasted with increases in Brittany and only slight reductions in Maine and Beauce; the southern fringe of the Massif Central underwent only a small decline, unlike the intense losses in Languedoc; while four départements adjacent to Gironde (down 68 600 ha) managed to increase their ploughland between the 1830s and 1907. In addition, the direction and tempo of change in individual départements proved to be remarkably complicated during successive cadastral phases.

At the time of the July Monarchy only 13 départements, in Languedoc, Lower Normandy, around the Gironde and in some upland regions, registered a decline in their arable surface. Over most of the country ploughland expanded at the expense of other realms, but only in Loire-Inférieure and Vendée did it increase by more than 1000 ha p.a., although annual advances of 500–1000 ha occurred in nine other départements along the fringe of Armorica, in middle France and in Picardy (Fig. 5.2a). Wasteland retreated in the face of the plough in western and middle France, while woodland was the main element to decline in Picardy and a scatter of other areas. The pays described by Lullin de Châteauvieux (1843, p. 77) as forming the *province des landes et bruyères* bore the brunt of this transformation, as the vast ploughlands of the Paris Basin were extended outwards into middle France. During the three decades that followed mid-century, only 49 départements increased their arable total and substantial losses were registered in Normandy, the north-east, the Alps and much of central and southern France. Nonetheless, large stretches continued to be tamed for cultivation in Armorica, middle France, and Picardy, and smaller examples of land being converted to arable use were found in Languedoc, the lower Rhône and the Ile-de-France. Over 1000 ha of new arable land were being added each year in 10 départements located in eastern Brittany, middle France and Languedoc; wasteland was the main element to retreat, except in the far south where phylloxerated vineyards were being ploughed up (Fig. 5.2b).

## Expansion and intensification

Nineteenth-century journals contain a vast but fragmented amount of information on ways of intensifying as well as extending arable cultivation, but very little was expressed in comprehensive or quantifiable terms. However, numerical data do exist on the processes of marling, liming, fertilising and underdraining during one decade of the Second Empire (Clout & Phillips 1972). The virtue of using calcareous materials to reduce soil acidity and to impart body had been appreciated in some pays for centuries, and was certainly mentioned by many administrators in the enquête agricole of 1814. Three distinct practices, each making use of different geological resources, were cited. Thus calcareous seasand was dredged from the shores of Armorica and applied to arable land close to the coast; marl pits pockmarked the landscapes of parts of the Paris Basin; and lime kilns functioned in Anjou and Maine following the discovery of small anthracite deposits at the dawn of the century (Risler 1897, Jardin & Tudesq 1972). Marl could be brought to the surface and applied to overlying clays with relative ease in much of the Paris Basin, as in the progressive districts of Seine-et-Oise where hauling marl up from as much as 10 m beneath the ground had been practised for several centuries (Richardson 1877). In the early 1880s most owner-occupiers in that département marled their land regularly, as did tenants on leases of nine years or more, but those on short leases did not make use of marl, for although its beneficial effects could last for 12–18 years they were not normally experienced until three or four years after application. Practices were far from uniform in neighbouring Beauce, where some farmers marled regularly but others were forbidden to do so in their leases. Further south in Indre it was claimed that marling had been practised from time immemorial and the agronomist Moll (1838–9) believed that it was essential for maintaining soil fertility in that département. Certainly the technique had made great progress since 1800 on the margins of the Sologne, and similar improvements were reported in several districts of southwestern France (Fabre 1829, Inspecteurs: Aude 1847, Haute-Garonne 1843, Hautes-Pyrénées 1843, Tarn 1845).

Liming of farmland close to kilns was recorded in several localities on the margins of the Paris Basin and in southern France during the early 19th century. For example, in Calvados lime was being applied successfully to fields alongside highways that served lime kilns even though haulage was causing serious damage to the roads (De Magneville 1827). However, to move from this local scale of activity and to attempt to transport lime long distances to the granitic regions of Armorica or the Massif Central, or to the outwash sands in middle France proved virtually impossible prior to the construction of the railways (Hoslin 1850). Indeed, the appalling quality of most rural roads, especially those in clayey, low-lying localities, provided even more hindrances. Nonetheless, many contemporaries were aware of the advantages that liming could impart and agricultural inspectors recommended that lime be applied to poor soils in many districts including the Montagne Noire in Aude (Inspecteurs: Aude 1847). Successive improvements to communications, such as the Nantes–Brest canal and the strategic roads of north-west France, proved most useful for moving lime and fertilisers and, in later years, every programme for taming mauvais pays during the Second Empire included schemes for improving communications. During the final decades of the 19th century, the railways were to come to play a vital role for feeding lime and fertilisers into the heart of Limousin and western Brittany, where new fields were still being created.

Detailed information on marling and liming was sought in 1862 and estimates were returned in 1873. Marling was recorded over 650 105 ha between 1852 and 1862 and occurred mainly in the Paris Basin, where calcareous substances underlay surface deposits in many localities and farming was prosperous enough to support the capital expenditure that marling required (Fig. 5.1b). In 10 départements over 20 000 ha were marled in that decade, with peaks in Eure (67 385 ha) and Eure-et-Loir (98 285 ha). These northern districts contained very little wasteland at this time and most marling must have served to enhance the fertility of land already in cultivation. In addition, quite extensive marling was taking place in parts of middle France (Loiret 31 780 ha, Loir-et-Cher 19 475 ha, Indre 15 120 ha, Cher 13 690 ha), where défrichement was widespread. In typical fashion marling was deemed to be essential if the Sologne were to be brought into cultivation, and a case for reducing the cost of carrying marl by rail was being prepared in the early 1860s (Ministère de l'Agriculture 1865–6).

The distribution of liming activity was quite different, with virtually none occurring in the Paris Basin proper but a great deal being practised along its southern and western margins, where suitable deposits of limestone could be exploited (Fig. 5.1c). Ten départements declared more than 20 000 ha being limed between 1852 and 1862 and most activity took place on the eastern margins of Armorica and in Bourbonnais-Nivernais, with peak values being returned for Mayenne (140 040 ha), Allier (55 795 ha) and Calvados (51 105 ha) (Dufour 1981, Martin & Martenot 1909). The émigration intérieure of 1830 heralded a phase of improvement in Maine, so that in 1846 the deputy of Mayenne could claim that liming had 'quadrupled the productivity of the soil, especially between Sillé-le-Guillaume, Laval and Sablé' (cited in Suret-Canale 1958, p. 303). Excavating limestone and anthracite provided additional rural employment, more land was put under the plough, cultivation of wheat advanced in place of rye and buckwheat (giving the département a marketable surplus), and growing artificial meadows on improved soils enabled many more livestock to be raised (Labrousse 1956). During the Second Empire extensive liming was practised on outwash sands in Bourbonnais and on granitic soils in Morvan, and produced an identical agricultural response to that in Maine (Chabot 1945, Derruau 1949). By contrast, very little liming was taking place in Limousin in the 1850s, although road improvements since 1836 did enable small quantities to be carted from Angoumois and Poitou (Perpillou 1940).

The only other quantitative evidence on marling and liming dates from 1873 and simply indicates the volumes of calcareous material that were being applied in each département. More than 2 000 000 qx (quintaux; 1 quintal equals 100 kg) of marl were being used in 10 départements, mainly in the Paris Basin, and the practice would seem to have extended in three additional

areas during the course of the 1860s. An important new focus had emerged in
Manche (3 690 000 qx), with Ille-et-Vilaine, Cher, Creuse and Indre standing
out more strongly in response to interior colonisation and improvements in
road and rail communication. For example, liming had been practised in a
small way in northern Limousin since the July Monarchy, using substances
from Vienne and Indre, but costs had proved prohibitive for all but the small
agricultural élite. Construction of main lines between Limoges and Brive
and towards Bordeaux during the Second Empire changed all this, enabling
lime to be brought in cheaply and stimulating bands of défrichement close to
the railway tracks (Perpillou 1940). Between 1880 and 1912 the pattern was
to change once again as minor lines were built which allowed lime to be
transported from kilns in the Charentes and Dordogne in the west, from
Montluçon and Le Châtre in the north, and from the basin of Brive (Corbin
1975, Dessalles 1937). A very similar trend was to occur further south
around Carmaux where local deposits of limestone and coal had been used to
make lime on a limited scale in the early 1800s and the substance had been
applied to land around Rodez, in the Ségala, and the Maurs basin. Only a few
farmers had been able to afford it but they 'did not know how to praise it too
highly' (Anon. 1867, p. 19). Transport costs remained great in this part of
central France and, as in the mauvais pays of the lowlands, a special case was
made for allocations of finance to enable agricultural roads to be built for
transporting lime. During the Third Republic road and rail were to be used
in concert to allow liming to advance across the Ségala, but in some remote
localities the practice was not started until the Rodez–Carmaux railway was
completed in 1902 or lorries began to be used after World War I (Cavaillès
1946, Taillefer 1978).

Little of a comprehensive nature is known about the use of other types of
fertiliser during the Second Empire. Responses to the great agricultural
enquiry of the mid-1860s were in general agreement that too few livestock
were being kept in most parts of the country and supplies of manure were
insufficient. In order to remedy this deficiency many commentators
advocated that fallows should be abolished, more fodders grown, additional
animals raised, and greater attention be paid to urban wastes as sources of
fertiliser. At this time above-average quantities of animal manure were
applied throughout the Paris Basin, Lower Normandy and the lower Rhône,
with Seine (452 qx/ha), Aisne (368 qx/ha) and five other northern
départements receiving more than 300 qx/ha (Puvis 1840–1, Barral 1852)
(Fig. 5.1d). By contrast, the intensity of manuring was well below the
national mean of 192 qx/ha in Burgundy, much of the south-west and the
Massif Central. The use of stable manure, green wastes, guano and 'other
commercial fertilisers' was recorded in 1873 and this information confirmed
the clear superiority of the Paris Basin. In addition, cavalry garrisons in the
Ile-de-France generated great quantities of horse manure and also provided a
ready market for locally grown supplies of fodder (Mollat 1971). The pattern

of fertilisation in the 1870s was very similar to that recorded a decade earlier, with more than 25 000 000 qx of farm manure being applied in each of a dozen départements arranged in an uninterrupted block from Nord to Loir-et-Cher and from Eure to Marne, and no less than 48 000 000 qx being used in Aisne. Application of natural fertilisers declined sharply beyond these areas.

Peruvian guano had first been brought into France at mid-century and imports increased rapidly thereafter, to reach 51 000 000 kg in 1857 and 57 000 000 kg in 1866 (Augé-Laribé 1955). The use of this fertiliser was mentioned in no less than 74 départements in 1873, but very small quantities were normally involved since most farmers were reluctant to use it for fear of adulteration or fraud, as well as for reasons of cost (Ministère de l'Agriculture 1865–6). Nord made use of a total of 940 000 qx, far ahead of Seine-et-Marne (610 000 qx), Oise (620 000 qx) and Seine-et-Oise (260 000 qx), with peak intensities of application involving farms close to railway stations or navigable waterways (Brunet 1960). Sizeable quantities were also being employed in Eure-et-Loir, Calvados and Vaucluse, each of which contained important examples of commercial farming at this time. Not surprisingly, guano was unknown in much of the Massif Central and Pyrenees. The 'other commercial fertilisers' were not itemised but must have included oil cake, noir animal, Chilean nitrates of soda, phosphates and sulphate of ammonia (Baud 1932). They were employed almost exclusively in the immediate environs of Paris (13 800 000 qx in Seine-et-Oise) which contained the richest farmers with capital to spend. Consumption in other leading départements (e.g. Nord 2 100 000 qx) was of a much more modest order of magnitude. Chemical fertilisers were either unknown or were used on a purely experimental basis over the greater part of France in 1873 (Brunhes 1900). In the 1880s they were beginning to be used quite widely in middle France but were only to become available at the turn of the century in the Ségala (Chavard 1937, Meynier 1931).

Better management of soil moisture formed a further essential component in the extension and intensification of arable activity, involving removal of surplus water in low-lying or clayey areas and provision of extra supplies in arid districts (Barral 1853). During the July Monarchy excessively damp ploughlands were being drained by surface ditches in many parts of the land but, in the mid-1860s, piped underdrains started to be installed in emulation of successful achievements across the Channel (Naville 1850, Phillips & Clout 1970). A pipe-making machine had been imported from Britain into Manche as early as 1839 but the real focus of activity was to develop slightly later on estates located on the stiff clays of Brie, where 'water had been known tó stand . . . after heavy rains, deep enough to float a boat' (Richardson 1877, p. 400). At mid-century Barral (1856) estimated that 12 000 000 ha of French farmland needed some kind of drainage and, between 1852 and 1862, 10 835 ha were duly provided with open drains. Most work was accomplished in the northwestern quarter of the country and in the

Garonne valley, although the largest individual totals were in Loir-et-Cher
(7795 ha) and Côte-d'Or (7530 ha). Unfortunately it is not possible to
determine what proportions of that total involved land under arable or in
pastoral use.

Large landowners in Seine-et-Marne had started to replace open ditches
with underdrains during 1846 and work began in neighbouring Seine-et-
Oise four years later (Bernard 1953, Evrard 1923). In 1854 legislation
established the right of landowners to pipe surplus water beneath their
neighbours' plots to reach the nearest stream and in the same year pipe-
making machines were sent to many provincial agricultural societies. By
1856 almost 400 had been distributed and 10 départements had more than
10 machines apiece, with the largest totals being in Seine-et-Marne (22), Ain
(19), Nord (14) and Pas-de-Calais (13). Some 35 000 ha had been drained by
the end of that year, with peak values in Seine-et-Marne (8000 ha) and Pas-
de-Calais (5000 ha). Following legislation in 1856 and 1858, the government
set aside 100 000 000 fr to loan to landowners who contemplated this improvement, but numerous administrative and legal formalities reduced the
utility of these well intended moves.

Some 118 685 ha were equipped with underdrains between 1852 and 1862,
with all départements in the nation mentioning some activity but Nord
(23 025 ha), Seine-et-Marne (10 895 ha) and Pas-de-Calais (11 140 ha)
accounting for two-fifths of the total. These three départements embraced
the deep claylands of Flanders, Brie and Boulonnais; contained important
foci of progressive agriculture; and were close to either Paris or England,
from which this particular technical innovation was being diffused. Thus,
agricultural societies in Nord and Pas-de-Calais had managed to acquire
machines directly from English suppliers and, as a result, landowners in
Boulonnais and Flanders were starting to adopt the technique by the early
1850s (Hubscher 1979–80, Tribondeau 1937). However, the land underdrained during that decade amounted to only 1 per cent of Barral's estimate
of what was needed, and the figure of 200 000 ha cited in the agricultural
enquiry of the mid-1860s was still of minimal significance. Such a limited
response was explained by many factors, including fragmentation of landownership and great morcellement in many parts of the country; sheer
expense, which was not really alleviated by the complex system of loans;
technical difficulties for ensuring adequate runoff; proliferation of short
leases, which disinclined tenants to press for improvements; and downright
indifference on the part of many landowners.

## Retreat

During the years from 1879 to the outbreak of World War I, ploughland
increased in only 16 départements, with more than 1000 ha of additional

arable being recorded in just six areas located in western Brittany, middle France and Aquitaine. This, albeit limited, advance of the arable realm largely reflected the final stage in the complex and fluctuating process of défrichement that had been in operation for more than a century; while, in addition, some phylloxerated vineyards were being converted to arable use in southwestern France. However, much more striking was the trend for arable to decline during the Third Republic. In 42 départements more than 1000 ha of ploughland were converted to other uses each year; in eight of these areas the rate of conversion rose to 2000–3000 ha, and exceeded 3000 ha in five départements of Lower Normandy and Languedoc (Fig. 5.2d). Ploughland retreated over extensive areas of eastern France, with parts of Aquitaine, Lower Normandy and Maine forming westerly projections of this zone of decline. Permanent grass was the main compensating form of land use to advance in most of northern and central France at this time, although vineyards were being established at the expense of ploughland in Languedoc, and sizeable stretches of cultivated land were simply falling out of productive use in mountainous areas and parts of the north-east.

Little is known about underdraining during the two decades after 1862, but between 1882 and 1892 only 95 450 ha were declared to be thus improved and this figure was substantially less than that for the 1850s. By 1878, 12 000 ha had been underdrained in Seine-et-Marne, essentially on the large farms of Brie, but in the subsequent decade and a half that figure increased by only a fraction to reach 14 790 ha in 1892 (Bernard 1953). The first syndical association for underdraining was not established in Brie until 1908. Nord (9930 ha) remained in second position in the underdraining league during the 1880s and was followed by Allier (4920 ha) and Creuse (3240 ha). By 1892 piped underdrains had been installed on more than 2500 ha in many départements of northern and middle France, but the onset of agricultural depression meant that there was precious little cash to spare for land improvement except on very efficient farms or on externally financed estates. In short, underdraining simply failed to be adopted widely and the total area that was improved in this way during the second half of the 19th century was pathetically slight.

Toward the other extreme of the environmental scale, areas of land in several parts of France were being put under irrigation for the first time (Beaudoin 1891). Unfortunately, no figures were collected before 1882–92, when 31 335 ha of ploughland and 5520 ha of market gardens were declared to be freshly irrigated (Chavard 1900). Six départements from the Rhône to the Pyrenees accounted for 55 per cent of all new arable irrigation, with most activity in Bouches-du-Rhône (7315 ha) and Vaucluse (3610 ha) (Bethemont 1972). A second focus of irrigation involved six départements in middle France and Limousin, where the practice was also long established.

Evolution of the arable realm between the 1830s and the eve of World War I conformed to the national trend across almost half of France (32 départements), with a phase of decline after 1879 following two phases of advance (+ + −) (Fig. 5.1e). This chronological model fitted a remarkable diversity of milieux, from great granaries in the Ile-de-France to *pays viticoles* in the Midi and many depopulating upland districts. Much of the Paris Basin, the fringes of Armorica, the north-east, Provence, the far south-west and the southern margins of the Massif Central evolved in a comparable way, even though the reasons which produced that common trend varied substantially from one locality to another. A quarter of the country (20 départements) displayed a retreat from arable activity following mid-century (+ − −) and such areas were especially widespread in Normandy and eastern France. In more extreme fashion, arable declined in seven départements during the whole period (− − −), with precocious development of pastoralism (e.g. Calvados, Cantal) and rising importance of viticulture (e.g. Aude, Gironde, Pyrénées-Orientales) making formidable contributions to this trend. Arable increased consistently (+ + +) in 13 départements located in Brittany and middle France, that formed the real agricultural 'frontiers' of 19th-century France, where cultivation continued to occupy more territory even during the agricultural depression. After an early phase of decline, Finistère, Charente and Loir-et-Cher increased their arable land after mid-century (− + +) and may be considered as extensions of the previous model, which thereby served to fill out the interior colonies of middle and north-western France. Only seven départements failed to conform to the preceding five patterns of arable change. Since each was heavily committed to viti-culture, the sharply changing fortunes of that activity may well have accounted for these areas' idiosyncratic evolution. For example, in Hérault the surface under the plough declined during the viticultural revolution prior to mid-century, increased substantially between 1851 and 1879 as the impact of phylloxera was felt, but then retreated as the reconstitution of vineyards took place (− + −). The collapse of vine growing may account for the late extension of arable land in Charente-Inférieure and Gers (− − +), while planting of new vines helps explain the interruption of arable extension in a number of relatively phylloxera-free départements in the valleys of the Garonne and Loire during the Second Empire (+ − +). Of course trans-formations such as these represented only one dimension of change to affect the arable realm; notable modifications were also at work with respect to individual crops and farming practices.

# 6    *Tradition and innovation*

## An initial view

In order to investigate changes within the arable realm it is necessary to turn
from the cadastre to the results of successive agricultural enquiries which cite
an arable total of 24 523 910 ha in the late 1830s and 23 194 840 ha in 1912
(Table 4.2). Definitions of ploughland were clearly not identical in the two
series of data but global figures were similar and convey the same pattern of
change through time. The implications of modified rotations, changing
spatial relations, innovations in land management, and new implements and
machines combined to engender substantial changes within the arable realm
in the century from Waterloo to World War I. Most notable was the retreat
of bare fallow, which had still formed the largest component of the arable
realm in the 1830s. Then came the advance of wheat, as poor soils were
steadily improved and as market demands spread into what had been semi-
subsistence territory. As a result, the noble cereal replaced secondary grains
which had flourished in isolated areas and at earlier times. However, freshly
cleared soil often proved unsuitable for wheat right at the beginning of
cultivation and hence rye, buckwheat and other rustic crops were grown as it
was gradually being improved. Fodder crops were sown more widely
between 1815 and 1914, which allowed important progress in animal
husbandry, and in addition a range of new crops, such as sugar beet and field-
grown vegetables, made an important appearance in a limited number of
localities.

No less than 27 per cent of French arable land was being left fallow in the
1830s but there were profound regional variations associated with different
rotations and the degree of acceptance of fodder crops (Duffoure-Bazin
1840). Over 40 per cent of ploughland lay fallow in eastern Armorica,
middle France, the southern Massif Central, the Camargue and many
mountain areas, with peak values in Aveyron (45 per cent), Allier (48 per
cent), Lozère (49 per cent) and Vienne (54 per cent) (Fig. 6.1a). Proportions
in excess of 35 per cent reflected either the periodic cropping and long
fallowing of poorish soils or the operation of degraded forms of biennial
rotation. Fallows occupied substantially less land in northern France but
values of 25–30 per cent showed that the triennial rotation survived
relatively intact in Champagne and the north-east (Dezeimeris 1841–2). Less
than one-fifth of ploughland was left fallow in a scatter of départements in
Normandy, the far north, Ile-de-France and the south-west, where fodder
crops (or maize in the case of Aquitaine) were sown on what had traditionally

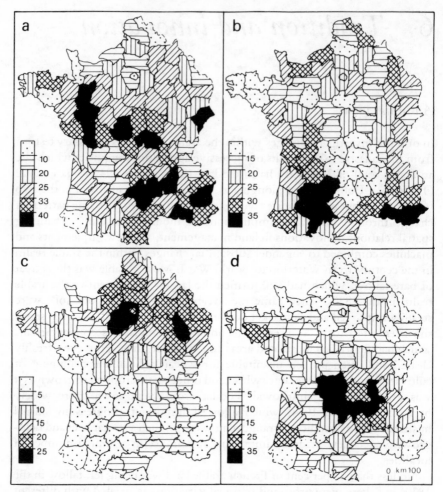

**Figure 6.1**  (a) Fallow, (b) wheat, (c) oats and (d) rye as percentage of arable land, 1837.

been the fallow portion. Bare fallow had almost disappeared in Nord (10 per cent), Basses-Pyrénées (8 per cent) and Manche (8 per cent).

Wheat was by far the most widely grown cereal in the 1830s and since 1815 its cultivation had spread on to a further million hectares (up 21 per cent) (Fig. 6.2). Rates of increase were most impressive in areas where other grains formerly had been grown; hence the wheat surface advanced by over 50 per cent in 18 départements of southern and western France and had doubled in six (Noirot 1838). The 5 383 235 ha under wheat covered 22 per cent of French ploughland in the late 1830s, but in southern localities with biennial rotations the proportion was much greater (e.g. Var 42 per cent, Gers 45 per cent, Lot-et-Garonne 48 per cent) (Fig. 6.1b). Ten other départements in the

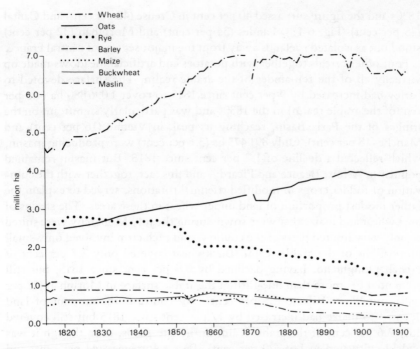

**Figure 6.2**   Evolution of surfaces devoted to cereals, 1815–1912.

Midi, Normandy and Nord devoted between one-third and two-fifths of their land to wheat, but in the Massif Central the proportion fell below 10 per cent and was exceedingly slight in Haute-Loire (4.1 per cent), Cantal (3.0 per cent) and Creuse (0.5 per cent).

Oats covered 2 992 365 ha in 1837 (12 per cent of national ploughland) and the amount devoted to this crop had increased by almost 500 000 ha (up 20 per cent) in the preceding two decades. Oats, of course, was long established in the triennial rotations of northern France and hence fastest rates of recent adoption were in southern, western and central districts. Land under oats had increased by more than 50 per cent in 34 départements and had doubled in 19 of these, but the Paris Basin remained the major focus of production nonetheless, with 14 départements having over one-fifth of their arable under oats and the proportion exceeding 25 per cent in Haute-Marne, Seine-et-Oise and Eure-et-Loir (Fig. 6.1c). Rye occupied a similar amount of land (2 537 645 ha, 10 per cent of arable) but displayed a very different distribution from oats and had declined by 0.7 per cent since 1815. In fact, strongly contrasting trends had been in operation, with rye retreating in 36 départements but advancing in 46 others, especially where fresh land was being brought into cultivation. Thus in 10 départements the amount of land under rye increased by over 50 per cent in just 20 years. More than a third of the ploughland in six départements in the Massif Central was devoted to growing rye in the late

1830s and the figure surpassed 40 per cent in Creuse (42 per cent) and Cantal (45 per cent) (Fig. 6.1d). Landes (34 per cent) and Morbihan (32 per cent) stood out as isolated ryelands away from the major ségalas of central France.

Four other cereals together with potatoes and artificial meadows made up virtually all of the remainder of the arable realm. The surface devoted to barley had increased by 9 per cent since 1815 to cover 1 096 880 ha (4.5 per cent of the arable realm) in the 1830s and was particularly significant on the fringes of the Paris Basin, reaching its peak in Vienne (16 per cent) and Manche (18 per cent). Only 891 475 ha (3.6 per cent) were producing maslin, which reflected a decline of 1.2 per cent since 1815. But maslin remained popular in Artois, Beauce and Picardy, and this fact, together with the cultivation of fodder crops in modified triennial rotations, served to explain the rather modest proportion of land under wheat in these areas. The strains of buckwheat and maize that were sown during the July Monarchy were suited to only very limited parts of the country, and each crop involved only small proportions of the arable realm. Buckwheat covered only 2.7 per cent of French ploughland, having declined by 0.5 per cent since 1815, but still accounted for more than one-fifth of the arable surface in Morbihan (22 per cent) and Ille-et-Vilaine (25 per cent). By contrast, the amount of land devoted to maize had increased by 17 per cent since 1815 but still covered only 2.6 per cent of the arable realm in the late 1830s, even though it was widely cultivated in Lot (20 per cent), Basses-Pyrénées (44 per cent) and Landes (57 per cent).

The range of crops known as artificial meadows had experienced varying histories of diffusion and were suited to quite different environments, so that each must have demonstrated a distinctive pattern of production during the July Monarchy (Chorley 1981). For example, clover grew best in lime-rich soils in cool, moist environments; lucerne required rich, permeable soils with a good humus content; while sainfoin flourished in many types of soil and was suited to southern France as well as the cooler north. Unfortunately, it is not possible to quantify the distribution of each of these crops, but collectively they occupied 6.2 per cent of the arable realm in the late 1830s and were widespread in the Ile-de-France, Picardy, Flanders and Normandy, where traditional rotations had been modified to incorporate them. Over one-eighth of the arable land in seven départements in the Paris Basin was devoted to artificial meadows (e.g. Oise 15 per cent, Seine-Inférieure 16 per cent) and they were also grown widely in the pays charentais, Languedoc and along the eastern fringe of France. Potatoes provided a vital adjunct to traditional food supplies but occupied a mere 846 840 ha (3.5 per cent) in the late 1830s and displayed a remarkably uneven pattern of cultivation. Only seven départements had over 8 per cent of their arable land devoted to them and these areas were located in the Massif Central, Périgord, the Vosges, central Pyrenees (Ariège, 13 per cent) and the environs of Paris (Seine, 12 per cent).

The Statistique of the July Monarchy provided immensely detailed information on the production and productivity of individual crops and such evidence has been examined elsewhere (Clout 1980). Wheat productivity was found to offer a reasonable guide to variations in the yield of other cereals and hence attention will be focused on wheat alone in the present discussion. But rather than rely solely on information from the 1830s, data from the next agricultural enquiry (1852) will also be introduced. By the middle years of the 19th century a fraction over 2.0 hl of seed was, on average, being applied to each hectare of wheatland; but there were of course very substantial deviations from this mean, with the sower's hand tending to be heavy throughout northern France and certainly being light in Aquitaine. During the unquestionably harsh years of 1815/20 wheat had given an average crude yield of only 9.88 hl/ha, but by the late 1830s this had increased to 12.45 hl and rose to 13.64 hl in 1852, giving a mean yield of 13.04 hl/ha for the last two dates. For France as a whole, crude wheat yields had risen by one-third from 1815/20 to 1837/52, with rates of increase exceeding 50 per cent in Champagne, Provence, Charentais and the pays of the Garonne, although absolute yields were not high in these regions. Rates of increase proved to be far more modest in progressive high-yielding areas, such as Nord and the Ile-de-France, and in isolated, backward regions, such as the Massif Central and parts of Armorica, where absolute yields were low and bread grains other than wheat tended to predominate.

The geography of crude wheat yields for 1837/52 was very similar to that for 1815/20 (Fig. 6.3a). Seine département yielded an impressive 23.7 hl/ha toward the middle of the century (some 80 per cent above the national average), being followed by Nord (21.8 hl/ha), Seine-et-Oise (20.5 hl/ha), Oise (19.4 hl/ha), Seine-et-Marne (19.3 hl/ha), Somme (19.2 hl/ha) and Seine-Inférieure (19.1 hl/ha). The traditional high productivity of well fertilised and deeply ploughed northern sections of the Paris Basin remained unrivalled (Dupâquier 1972). With the exception of Upper Brittany and western Lorraine, almost the whole of the northern third of France registered above-average wheat yields (Goubert 1957). At the other extreme, crude wheat yields were below 10.0 hl/ha in parts of the Alps, the Massif Central and Poitou, plummetting below 8.0 hl/ha in Dordogne (7.94 hl/ha) and Lot (7.20 hl/ha). Unfortunately, crude yields are a rather misleading indicator of productivity, since they disregard the volume of seed that was applied, and hence seed : yield ratios have been calculated as they provide a more reliable guide. The average seed : yield ratio for wheat in 1837/52 was 1 : 6.27 and only in Nord (11.08), Pas-de-Calais (9.70), Somme (8.78) and Seine-et-Marne (8.43) did it exceed 1 : 8.0 (Fig. 6.3b). From this calculation it is clear that the high crude yields of Seine and Seine-et-Oise were partly a reflection of the sower's heavy hand. Below-average ratios characterised the whole of France beyond the Paris Basin and descended to desperately low levels in Cantal (3.83) and Lozère (3.67). The deeply ploughed, well manured, inherently fertile arable land of the Ile-de-France, Flanders and the northern Paris Basin reigned supreme.

## Trends of change

The 10 main components of the arable realm displayed substantial and varied changes over the full span of years from the July Monarchy to World War I. Wheat, oats, artificial meadows and potatoes each tended to occupy progressively larger areas nationwide, while fallows and the five remaining cereals tended to retreat (Table 6.1). However, these basic trends were not consistent either through time or across space. For example, the amount of land devoted to wheat was 20 per cent greater in 1912 than it had been in the late 1830s but the maximum surface had actually been reached in 1882 (31 per cent greater than 1837) and the national total declined during 1862–73, 1882–92, and 1902–12 in response to the widespread ramifications of the agricultural depression and the conversion of ploughland to other land-use realms. In addition, there were very profound variations in the combinations of cereal crops that were grown in individual départements. Cultivation of wheat declined in many localities because of land abandonment or changing emphasis within the local rural economy, but in other départements more territory was devoted to this crop as the soil was improved through marling, liming and fertilising and as the production of other grain retreated (Vilmorin 1880). For example, a 'veritable agricultural revolution' occurred on the plateau of Limoges between 1860 and 1890, as wheat replaced rye across 30 000 ha of ploughland (Perpillou 1927, p. 512).

Simple comparison of land use in the late 1830s and 1912 shows that wheat cultivation had advanced to occupy more land in 58 départements, but had declined in the remaining 24 (Fig. 6.3c). But to make such a direct comparison is somewhat misleading, since only 16 départements (located in the

**Table 6.1** Peak cereal surface (départements).

|  | 1837 | 1852 | 1862 | 1873 | 1882 | 1892 | 1902 | 1912 |
|---|---|---|---|---|---|---|---|---|
| wheat | 0 | 5 | 29 | 7 | 9 | 15 | 1 | 16 |
| oats | 3 | 9 | 3 | 1 | 6 | 12 | 5 | 43 |
| artificial meadows | 1 | 4 | 11 | 10 | 5 | 19 | 2 | 30 |
| potatoes | 2 | 0 | 10 | 4 | 9 | 17 | 5 | 35 |
| fallows | 72 | 10 | 0 | 0 | 0 | 0 | n.d. | n.d. |
| rye | 66 | 12 | 0 | 3 | 1 | 0 | 0 | 0 |
| maslin | 60 | 4 | 7 | 3 | 4 | 2 | 0 | 2 |
| barley | 37 | 4 | 2 | 12 | 12 | 3 | 0 | 10 |
| buckwheat* | 17 | 21 | 9 | 0 | 8 | 5 | 2 | 2 |
| maize* | 7 | 4 | 2 | 12 | 12 | 3 | 0 | 10 |

*Not present in every département.

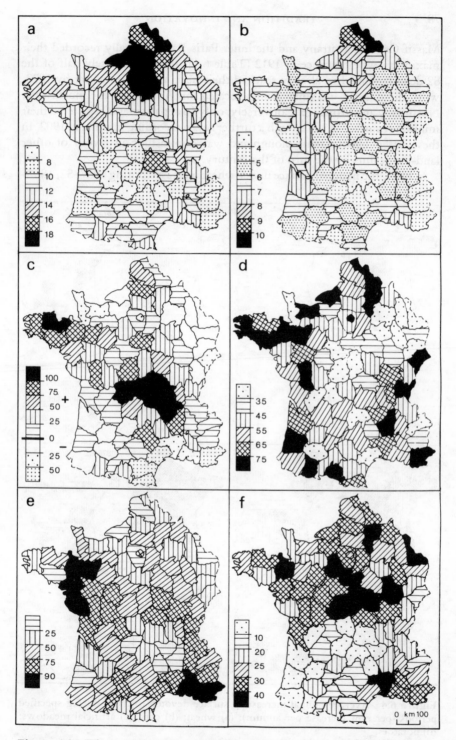

**Figure 6.3** Wheat: (a) average yields, 1837/52 (hl/ha); (b) seed : yield ratios, 1837/52; (c) net change in surface, 1837–1912 (per cent). (d) Fallow, reduction in surface, 1837–92 (per cent). (e) Rye, net decrease in surface, 1837–1912 (per cent). (f) Improved ploughs as percentage of total, 1862.

Massif Central, Brittany and the inner Paris Basin) actually recorded their maximum wheat surface in 1912 (Table 6.1). Indeed, precisely half of the 82 départements had already reached their peak area under wheat by 1873, with these early maxima being located in Aquitaine, the north-east, Normandy and Provence. Conversely, only 16 départements registered their minimum surface in 1912, with a further seven recording minima in 1902. In these latter départements ploughland was diminishing in favour of other land-use realms by the turn of the century.

In brief, the national surface under wheat increased during 1837–52, 1852–

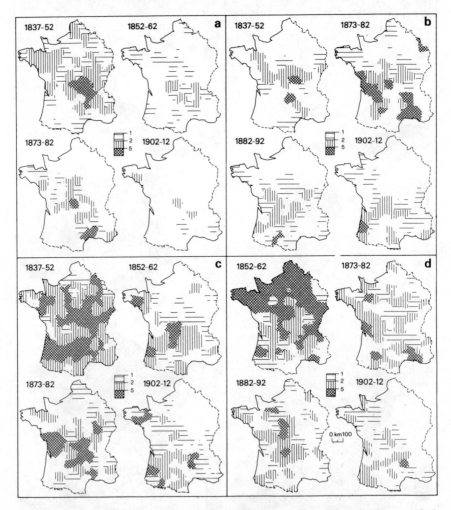

**Figure 6.4** Average annual increase in surface devoted to crops during specified periods (per cent increase per annum): (a) wheat; (b) oats; (c) artificial meadows; (d) potatoes.

62, 1873–82 and 1902–12 (Fig. 6.4a). During the July Monarchy the amount of land being sown with wheat was rising by more than 1 per cent each year in 57 départements, with very important advances in Armorica, Burgundy, Picardy and the Massif Central. Indeed, in the ségalas at the heart of the Massif the amount of land under wheat was increasing by more than 5 per cent per annum. Thereafter, the national advance of wheat cultivation decelerated or even reversed in each successive decade, with 23 départements advancing by more than 1 per cent per annum in the 1850s, 22 during the 1870s but a mere five after 1902, when Creuse, Haute-Loire and Morbihan were the only départements to extend their wheatlands by more than 2 per cent per annum. The total output of France's major bread grain rose substantially as a result of this trend, however the cultivation of wheat advanced at a rather slower rate than did oats, potatoes or artificial meadows.

The amount of land sown with oats increased by 1 031 985 ha between 1837 and 1912 and, because of this crop's utility as a fodder source, continued to do so not only during the agricultural crisis but indeed in every decade except 1862–73. As a result, and in striking contrast with the evolution of wheat cultivation, 43 départements recorded their maximum surface under oats in 1912. Cultivation of this crop continued to advance in Lower Normandy and other pastoral areas where wheat growing was on the decline; by contrast, the general retreat of arable husbandry in northeastern France also carried oats production in its train (Désert 1978). With the passage of time, cultivation of oats expanded from the Paris Basin outwards to the south and west as triennial rotations replaced biennial systems and as this crop was sown on areas of newly cleared land. Such a progression also reflected growing demands for fodder to support greater numbers of livestock and especially farm horses. This trend permitted heavier ploughs to be used in areas where draught animals became more numerous. Between 1837 and 1852, 25 départements increased their surface under oats by more than 1 per cent per annum, with important advances in middle France, the Armorican marshlands, the northern fringe of the Massif Central, and in Languedoc (Fig. 6.4b). Extensive défrichement was occurring in many of these areas and oats cultivation advanced as regular rotations were starting to be employed. The crop continued to make moderate progress in middle France during the 1870s but made faster ground in the lower Rhône and parts of Aquitaine. During the 1880s it advanced by more than 1 per cent per annum in 29 départements, and continued to progress by that rate in exactly the same number of départements in both of the following decades.

The land devoted to artificial meadows increased from 1 531 545 ha in 1837 to 2 948 810 ha in 1912 (up 93 per cent) in spite of a slight decline during the 1890s. During the July Monarchy these crops had advanced by more than 1 per cent per annum in 77 départements, especially in areas where relatively little fodder had been sown previously (Fig. 6.4c). The pattern had changed substantially by the 1850s and was assuming the basic features that it was to continue to display until World War I. Rates of adoption were slight in the northern quarter of France but continued with vigour in western and central départements where arable farming in general and triennial rotations in particular were making important progress. Land devoted to artificial meadows increased by more than 1 per cent per annum in 37 départements during the 1850s but stabilised at 41 during the 1870s and at 39 after 1902. These crops achieved their maximum extent in 1912 in 30 départements located predominantly in western and southwestern France; but by contrast peak surfaces had been reached as early as 1862 in Flanders, Picardy and the Ile-de-France, where sugar beet, field vegetables and other crops had come to be grown much more widely by the end of the century.

From a rather modest starting position, the land under potatoes increased by 80 per cent between the 1830s and World War I, but the process was far from evenly spread through space and time. For example, potato cultivation collapsed temporarily and with disastrous effect in the late 1840s (761 200 ha in 1852, compared with 846 840 ha in 1837), while the surface recorded in 1892 was fractionally greater than that for 1902. Nonetheless, important expansion continued in some départements up until World War I, and 35 départements, predominantly in central and northwestern France, displayed their peak surfaces in 1912 (Table 6.1). By contrast, cultivation of this crop declined after 1882 in several clearly defined regions: (a) the north-east, where woodland advanced in the place of arable; (b) Normandy, where permanent grass was replacing ploughland; (c) Picardy and parts of the Ile-de-France, where more remunerative crops were being grown; and (d) in the Mediterranean south, where the arable realm fluctuated markedly in response to the changing fortunes of the vine. During the 1850s potato cultivation had advanced across very different areas from those being occupied by other crops (Fig. 6.4d). Some 72 départements increased the surface under potatoes by more than 1 per cent each year, with rates of adoption being greatest in northern districts, but once that initial burst of acceptance had been achieved the advance of potato cultivation proceeded less widely during the 1870s (49 départements exceeding 1 per cent per annum) and came to focus on Armorica, middle France, the Massif Central and south-west in the 1880s (40 départements) and 1890s (27 départements).

Many other arable components contracted in varying degree between the July Monarchy and World War I. Central to all these changes was the retreat of bare fallow, which declined from 6 622 600 ha in the late 1830s to 3 083 940 ha in 1892 (down 53 per cent) (Fig. 6.3d). Unfortunately, no detailed returns were made subsequently. Some 58 départements recorded their minimum fallow surface in 1892 and contraction undoubtedly continued after then, with Augé-Laribé (1925) estimating that 2 400 000 ha were under fallow in 1912. From a national total of 2 537 645 ha in 1837 the amount of land under rye declined in every later decade to reach 1 186 550 in 1912 (down 53 per cent) (Fig. 6.3e). Its peak surface had been recorded in 1837 in

16 départements and occurred in 1852 in a further 12, located in mountainous areas and in middle France where rye was being used as a transitional crop between défrichement and ultimate cultivation of wheat. The amount of land producing maslin fell throughout the period from 891 475 ha in 1837 to 124 300 ha in 1912 (down 86 per cent), although peaks were recorded at the end of the century in Aveyron, Corrèze, Creuse and Côtes-du-Nord, which reflected the continuing advance of arable farming in these areas (Fig. 6.2). Buckwheat retreated by 29 per cent from 650 320 ha in 1837 to 459 680 ha in 1912 but retained its utility on poor soil and where land was being freshly tamed, as for example in Brittany. The decline of maize was of comparable dimensions, from 628 780 ha to 472 500 ha (down 25 per cent) but its pattern of cultivation was quite different, being grown essentially in the south-west, with a secondary focus in the Saône valley. Finally, barley contracted by 32 per cent from 1 096 880 ha to 749 700 ha, with peak surfaces occurring in 1837 in 37 départements. Only 10 départements in Armorica and along the northern fringes of the Massif Central devoted more land to barley in 1912 than early in the July Monarchy, since the general trend, at least until fairly late in the 19th century when new imperatives took over, had been to turn to wheat production wherever soils could be improved sufficiently (Zolla 1888).

Much less is known about the changing fortunes of minor arable components since these were recorded with much less rigour than the leading cereal crops. However, a sharp contrast may be discerned between traditional crops, which had been grown widely for domestic use during the ancien régime économique but subsequently fell out of favour, and relatively new specialised crops which were cultivated to feed growing numbers of livestock or urban dwellers (Peeters 1975). Thus, the surface under flax declined from 96 670 ha in the late 1830s to 27 750 ha in 1912, with only Nord (5935 ha) and Seine-Inférieure (6050 ha) supporting sizeable areas on the eve of World War I (Landureau 1883). Hemp underwent an even more drastic retreat, from 166 865 ha to 13 700 ha over the same period, with the single département of Sarthe (4800 ha) accounting for almost one-third of the latter total. Some 158 850 ha had been devoted to coleseed in the late 1830s, with nine northern départements supporting over 5000 ha apiece, which were headed by Nord (20 860 ha) and Pas-de-Calais (25 170 ha). Three-quarters of a century later, only 25 050 ha were grown throughout the country, with the modest peak value of 6340 ha being recorded in Seine-Inférieure. By contrast, market gardening and the production of various types of beet advanced dramatically from mid-century to World War I (Puvis 1843–4). In 1852, 35 060 ha of market gardens had been recorded, with leading concentrations in the environs of Paris (Seine-et-Oise 3675 ha), in the lower valleys of the Loire and Rhône, and in Flanders (Abeausy 1839–40, Jeantet-Maret 1940–1). Forty years later, flower nurseries and market gardens covered a combined surface of 141 570 ha and virtually doubled in the subsequent 20 years to reach 244 920 ha in 1912, with 13 départements containing more than 5000 ha apiece. Seine-et-Oise (12 680 ha) supplied only a modest share of the needs of the capital and fruit and vegetables were rushed to Paris from many parts of the nation, including Bouches-du-Rhône (10 380 ha), Gironde (11 180 ha), Dordogne (14 470 ha) and many other départements.

Beet of all kinds had covered 103 560 ha in 1852 and were grown in many départements, although Nord (24 430 ha) and Pas-de-Calais (13 800 ha) clearly led the field (Lerolle 1852). Sugar beet were listed specifically in 1862, when the national total reached 135 450 ha, with both Aisne and Nord surpassing 30 000 ha under sugar beet (Clout & Phillips 1973). By 1892 exactly twice as much land (271 160 ha) had been devoted to this crop, with a slight decline to 255 170 ha ensuing in the subsequent two decades. Five départements that covered Picardy, Artois and Flanders accounted for no less than four-fifths of the total, with the claylands of Brie (Seine-et-Marne 6 per cent) forming a less significant outlier close to the heart of the Ile-de-France. By the eve of World War I, some 56 880 ha were devoted to growing beet from which alcohol was distilled, and this crop displayed exactly the same northern distribution as sugar beet. The progress of fodder beet is known less clearly but, in 1902, 537 240 ha were under this crop, with strikingly different patterns of production emerging by comparison with other types of beet. No fewer than 24 départements devoted more than 10 000 ha apiece to this crop, with a very important concentration standing out in the pastoral départements of northwestern France (e.g. Ille-et-Vilaine 23 915 ha, Deux-Sèvres 24 555 ha) complementing the long-established stretch of beet-growing country which extended from Flanders through Picardy into the Ile-de-France (e.g. Seine-et-Marne 20 995 ha).

## Ploughing the fields

A fundamental contrast had long existed in France between the light southern *araire*, that could be manoeuvred around obstacles but only scratched the soils and therefore failed to dig in manure or remaining weeds, and the heavier northern *charrue* that was more suited to damp clayey soils, cut a deeper furrow and allowed the soil to be turned adequately (Bloch 1931, Dion 1934). Despite its lack of weight, the araire was effective on very small plots and on sloping ground and was also well suited to fragile environments where deep ploughing might bury thin fertile humus layers (Livet 1978). On very fragmented holdings and in particularly difficult terrain, ploughs were replaced by hoes or spades and such implements were certainly being 'held in great honour' in the Limagnes of Auvergne as late as the 1840s (Jusseraud 1841–2, p. 32). Precious little mention was made of improved ploughs in the enquête agricole of 1814 and only around Pontoise was great enthusiasm reported for new implements. Elsewhere tradition reigned supreme, with ploughs being pulled by horses in the north and by oxen, cows or mules in other parts of France (Haudricourt & Delamarre 1955). Indeed, the whole matter of ploughing technology was viewed with disgust by many contemporary writers, such as the 'recent traveller' cited by Picot de la Peyrouse (1819) who declared, 'in nothing is France so deficient as in agricultural implements' (p. 89).

At the end of the First Empire a number of armament workshops started to manufacture agricultural implements, but acceptance of innovation was to prove exceedingly protracted (Duby & Wallon 1976). The whole of the 19th century formed a time of experimentation and slow adoption of robust ploughs that cut and turned the soil, rather than merely scraping it. Draught animals were luxuries in many parts of France prior to mid-century but human labour was generally abundant, and this fact encouraged the continued use of simple implements (Manry 1974). Mathieu de Dombasle started to manufacture light iron-shod charrues in 1823 and use of such instruments diffused from his base at Roville near Nancy across the plains and plateaux of northern France. Likewise, some northern blacksmiths started to experiment with robust, modernised versions of ancient ploughs (Brunet 1960). As a result, iron-tipped ploughs were being accepted on a limited number of market-orientated holdings in the Paris Basin during the 1820s and 1830s, but even in such fertile pays as the cornlands of Picardy traditional implements remained dominant until the middle years of the July Monarchy (Fossier 1974).

Very few holdings in the south adopted improved implements at this stage, although some progress started to be made in the Rhône delta during the 1830s (Anon. 1839). New instruments, such as the Dombasle plough, made their first appearance rather later in other southern parts of France, with the 1850s being cited in Aveyron, Cantal and Tarn-et-Garonne, the 1880s in parts of the Alps, and the 1890s in Bresse and the eastern Massif Central (Séverin-Canal 1934, Boulmier 1951). A lag of at least half a century separated the technically progressive north and north-east from the backward centre and south. Popularisation of 'Brabant' or 'Belgian' ploughs came rather later but followed a roughly similar diffusion route. These instruments had been known in Soissonnais in the 1830s, in Berry after 1840, and were being acquired by 'even the most modest farmers' in Picardy in the 1850s, but they did not reach the Ségalas until the 1880s and parts of Auvergne, Limousin and the Alps until after 1900 (Brunet 1960, Fossier 1974, p. 368). First appearances were, of course, a completely different matter from widespread acceptance and in parts of Provence ancient araires were only just being set aside in favour of heavier ploughs in the early years of the 20th century (Livet 1962). Acceptance of Brabant ploughs did not start until after World War I in Haute-Vienne, and in southern parts of the Massif Central the araire reigned supreme until it was replaced by the Brabant in the 1930s (Nanton 1957–63).

Broad spatial trends such as these emanated from myriad centres of experimentation, as blacksmiths and engineers developed their own versions of pattern-book models. For example, workshops near Rennes, Quimperlé and Redon formed foci of innovation in Brittany during the Second Empire (Robert-Muller 1932). Special needs stimulated the development of special types of plough. Artisans during the 1850s made light but robust ploughs to work stony soils in Berry, and a distinctive *charrue de Brie* was devised that was particularly effective for digging out roots of artificial meadows. Cultivation of sugar beet yielded yet another spate of specialised implements which allowed deep ploughing. Hence during the final quarter of the 19th century the damp clays of the northern Paris Basin were starting to be worked to depths of 35–40 cm by new ploughs that were pulled by 6 to 10 pairs of oxen (Hitier 1901). Deeper ploughing enhanced crop yields almost everywhere that it was attempted and allowed the agronomic qualities of thick damp clays to be appreciated at the expense of drier, more easily worked soils. In addition, strong iron-shod ploughs were essential for breaking in former moors and heaths, and for this reason special défrichement ploughs were assembled in areas of widespread reclamation, such as Haute-Vienne during the Third Republic (Dessalles 1937).

Little of this complex transformation was chronicled in the agricultural enquiries, with no statement even on the number of ploughs being made until 1852, when 2 492 530 ploughs were recorded. The first set of descriptive statistics appeared 10 years later when *charrues du pays* were distinguished from 'improved' ploughs, which accounted for one-quarter of the total. Both categories undoubtedly embraced great internal diversity. Traditional local ploughs were particularly widespread in central and southwestern France, comprising over 90 per cent of ploughs in the Massif Central and Aquitaine. Improved ploughs were found mainly in northern France, although Provence and Languedoc formed a detached concentration (Fig. 6.3f). They were surprisingly insignificant in the Ile-de-France and the north-east, but exceeded half the total in four départements of middle France (Nièvre, Indre, Côte-d'Or 51 per cent each, Cher 61 per cent). This distribution bears a reasonable resemblance to the broad north–south distinction between charrues and araires (Delamarre & Hairy 1971). Improved ploughs with front wheels comprised 14.5 per cent of all ploughing implements in 1862 and were particularly numerous in northern départements, exceeding two-fifths of all improved instruments in the core of middle France, Ille-et-Vilaine, Meurthe and Eure-et-Loir.

Later enquiries employed a different definition and drew a distinction between single- and multiple-shared ploughs. The number of implements with two or more ploughshares increased from 156 135 in 1882 (5 per cent of the total) to 262 500 (7.3 per cent) in 1892. Single-shared ploughs were particularly important in the totals for Lower Normandy, Upper Brittany, middle France and Aquitaine, forming over 97.5 per cent in each area (Fig.

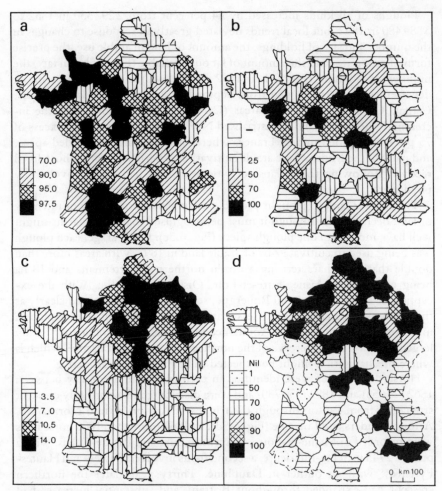

**Figure 6.5** (a) Single-shared ploughs as percentage of total, 1892; (b) percentage change in all ploughs, 1852–92; (c) arable surface (ha) per plough, 1892; (d) horses as percentage of draught animals, 1873.

6.5a). By contrast, multiple-shared ploughs comprised more than 30 per cent of implements in Picardy and adjacent areas (Aisne, 58 per cent; Oise, 61 per cent; Somme, 67 per cent), having been adopted on highly commercial farms that specialised in sugar beet and other market-orientated crops. Multiple-shared ploughs increased by 68 per cent during the 1880s, with numbers doubling or even tripling in parts of southern and western France, although absolute totals remained quite low. However, more than 3000 additional multiple-shared ploughs were recorded in four départements in Picardy in 1892, three along the Rhône valley, and in Creuse and Côtes-du-Nord, where extensive défrichement was taking place.

Ploughs of all kinds increased by 44 per cent from 2 292 530 in 1852 to 3 589 480 in 1892, but local trends deviated greatly in response to changes in the number and size of holdings, the amount of land in arable use, the precise form of cropping, and the amount of labour available (Fig. 6.5b). In fact, the number of ploughs recorded in 1892 fell below the 1852 figures in six départements, which ranged from depopulating highlands (Hautes-Alpes, Cantal, Corrèze) to pastoral areas (Calvados, Haute-Saône) and the increasingly urbanised and industrialised Nord. Rates of increase in excess of 75 per cent typified middle France, where land clearance proceeded apace, and parts of the Midi, where arable cultivation experienced some short-lived recovery following the collapse of viticulture that had been provoked by phylloxera.

The ratio of land to ploughs provides a convenient summary of intensity of land occupation, although it must be recognised that larger farms might well have more than one plough apiece (Fig. 6.5c). On average, each plough was being used to cultivate 7 ha of arable land in 1892, with areas more than double that size characterising a dozen northern départements and 18 ha being exceeded in Aisne, Eure-et-Loir, Oise and Meurthe. With the exception of Berry, Poitou and Rouergue, southern France emerged clearly as the realm of petite culture, with less than 3.5 ha of arable land to each plough in a dozen départements of Aquitaine, Dauphiné and Languedoc. Really large farms were the preserve of the northeastern quarter of France, which is where mechanisation might be expected to be particularly attractive.

Horse-drawn hoes were identified in each agricultural enquiry between 1862 (25 055) and 1892 (250 730) and were distributed in a northerly crescent stretching from Poitou through the fringes of Armorica, into Normandy, Artois, Flanders and Lorraine (Fig. 6.5d). In 1862 more than 1000 horse hoes were recorded in Maine-et-Loire, Mayenne, Eure and Pas-de-Calais, with a peak of 2745 in Seine-Inférieure; a second focus involved Basses- and Hautes-Pyrénées, with an outlier in Dauphiné. Thirty years later the northern crescent had expanded throughout Brittany and substantial increases had occurred in Picardy, and in central and southern parts of the Paris Basin. The southern crescent advanced through Languedoc and the lower Rhône to Isère, which had no fewer than 8385 horse hoes in 1892. The successive revival first of arable farming and then of viticulture toward the end of the century doubtless accounted for this trend in the far south.

## Sowing seed; harvesting grain

Cereals were harvested by hand on most farms throughout the 19th century but mechanisation was making a modest start on large holdings located in progressive areas. Agricultural returns included basic statistics and thus allow a broad picture of mechanisation to be drawn but this must be set in the context of spatial variations in topography, farm structure and availability of finance, labour and draught power (Grantham 1975). Mechanisation was inappropriate on small, labour-intensive holdings and required levels of investment that were simply impossible on many peasant farms. Early machines were not very robust and required level ground if they were to operate effectively (Walton 1979). Animals were needed to push or pull many of them but such draught power was lacking on many holdings. For these and other reasons mechanisation may be expected to appeal to relatively affluent farmers who worked fairly large holdings on level or gently sloping terrain (Winchester 1980). In addition, precocious decline in agricultural labour would provide a very real stimulus for technological innovation (Faucher 1954).

Threshing by flail was a particularly arduous task which demanded large labour inputs; however, depopulation and rising labour costs made mechanisation an attractive alternative for farmers who had capital to invest. For example, severe labour shortages were being experienced during the 1850s in Marne and in other départements and a wave of machinery purchase was the immediate result (Lamairesse 1861). As early as 1818, Mathieu de Dombasle had imported threshing machines from England and Sweden and had started to manufacture his own variety in the very same year (Jardin & Tudesq 1972). These machines made an appearance in the inner Paris Basin during the early 1830s but were still rare 10 years later, being displayed as curiosities at agricultural shows in many parts of the land (Drouyn de Lhuys 1875). They gradually increased in popularity during the Second Empire as French manufacturers turned out cheaper but less well finished versions of the machines that had been imported from across the Channel or the Atlantic (Richardson 1877). Such threshers operated in a stationary position, being turned by horses or mules; they were relatively simple in mechanical terms and did not have to be propelled across fields (Collins 1972–5). Hence it is not surprising that they formed the most numerous type of arable machine not only in 1852 but throughout the second half of the 19th century. Some 57 150 were recorded at mid-century but their relative importance declined somewhat from 85 per cent of all arable machines in 1862 to 72 per cent in 1892. Seven northeastern départements contained over 1500 threshers apiece in 1852, with peak totals in Meurthe (3055), Haute-Marne (4045) and Meuse (4880) (Fig. 6.6a). A second concentration was in the south-west, where Gers, Gironde and Haute-Garonne each had more than 1500, with an incipient focus on the margins of Armorica, where Maine-et-Loire, Sarthe and Mayenne contained over 1200 apiece.

The first and third foci still emerged clearly 10 years later but rather fewer threshing machines were recorded in Upper Languedoc. Ten northeastern départements each contained more than 2000, with their combined total having risen by 68 per cent from 22 860 to 38 425; however the fastest rates of adoption involved seven départements in the north-west, where the number

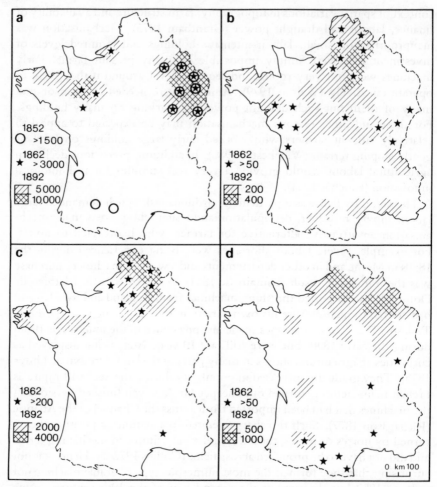

**Figure 6.6** Arable machinery: (a) threshing machines; (b) agricultural steam engines; (c) drills; (d) reaping machines.

of machines rose by 392 per cent from 4660 to 22950. Simple threshing machines were rare in the Ile-de-France in 1862, although Seine-Inférieure (3300) formed something of a solitary focus in nearby Upper Normandy. Very often owners would share the use of threshing devices with neighbouring farmers, and sometimes itinerants took more complicated versions of these machines from farm to farm, with local landowners providing the necessary complement of horses and men. The national rate of increase decelerated during the 1860s, accelerated in the 1870s, but slowed again in the 1880s to produce a total of 229750 in 1892. The foci in northeastern and northwestern France contained respectively two-fifths (91885) and one-quarter (61445) of the total, much the same proportions as in 1862. The greatest absolute increases in threshing machines occurred in these very regions, with some départements acquiring more than 6000 new machines, and Haute-Saône, Côtes-du-Nord and Finistère each containing over 9000 additional threshers. Half a dozen départements in Flanders, Picardy and the Ile-de-France acquired over 1500 new machines apiece, but such numbers were small by comparison with the total in the north-east and north-west. Threshing machines were virtually unknown in the Midi and, indeed, on the great majority of farms throughout the country where threshing by flail or under the hooves of horses or mules would appear to have continued as in former times (Meynier 1931).

In fact, the situation may have been somewhat different by the end of the century since large numbers of agricultural steam engines, either fixed or mobile were recorded in Picardy and at the core of the Paris Basin. They increased nationwide from 1520 in 1852 to 11 905 in 1892, operating a variety of barn machines and providing mobile sources of power for steam ploughs as well as threshers (Heuzé 1852). By mid-century, steam engines were being shown at fairs in many parts of the country but were numerous in only five départements of eastern France (Côte-d'Or 52, Haute-Saône 58, Jura 69, Saône-et-Loire 37, Doubs 26), which did not correspond precisely with the early focus of mechanised threshing in the north-east. Only a handful of steam engines were recorded in the Paris Basin, with Seine-et-Marne and Seine-et-Oise together containing only ten in 1852 (Bougeatre 1971). A curiously high total (900) was returned for Lot-et-Garonne, which must have been erroneous and perhaps the national total should be scaled down to take that into account. The pattern of adoption changed considerably during the 1850s and three clusters of départements with more than 50 steam engines apiece emerged by 1862 (Fig. 6.6b). Eight départements in the Ile-de-France, Picardy and Flanders together contained 670 (Nord 120, Aisne 115, Oise 100); four eastern départements contained 420 (e.g. Ain 175, Saône-et-Loire 100); and five départements on the margins of Armorica had a total of 535 steam engines (including Maine-et-Loire 125, and Vendée 165).

National trials of steam engines on large farms in the Ile-de-France served to increase their popularity among affluent farmers so that by 1892 not only had the national total risen to 11 905 but the relative importance of the various clusters of adoption had also changed (Le Gallais 1869). The number of engines in eastern and western France advanced modestly but the total in the Ile-de-France and Picardy rose by 342 per cent to 2960 machines (e.g. Seine-et-Marne 540, Aisne 515). Large numbers had also been introduced across the ploughlands of middle France, where 10 départements contained over 250 each, and a handful of southern départements had over 200 machines apiece (Hérault 310, Haute-Garonne 210, Bouches-du-Rhône 205) but some of these may well have been used for ploughing, pumping water or undertaking other functions as well as providing power for threshing. Certainly the value of steam ploughing to depths of 35–40 cm had been acknowledged in the Paris Basin, but the technique was not widespread even in the final decade of the century (Hitier 1902). Regardless of the acquisition of threshing machines and steam engines, the great bulk of threshing must still have been done by flail at the dawn of the new century, almost exclusively so in central and southern France and on small farms throughout the land.

Broadcasting undoubtedly remained the most widespread mode of sowing seed, even though drills had been used on an experimental basis in the inner Paris Basin as early as the 1830s, and 10 705 were in use nationwide in 1862 (Jardin & Tudesq 1972). Nonetheless, France contained some 5 500 000 farms in 1892 but only 52 305 drills, of which two-fifths were for sowing cereals, one-fifth for root crops, one-tenth for distributing fertilisers and the remainder for a combination of functions. To ensure even flows of seed offered many advantages: the crop would grow more evenly, mature more uniformly and yield a more homogeneous harvest. Evenly spread plants would receive more air and light than if they were bunched, and would support more robust stems which would prove more resistant to rot and disease. Use of a seed drill permitted a saving of 25–30 per cent in the quantity of seed applied and a comparable increase in yield might reasonably be expected (*Grande encyclopédie* n.d.). On the other hand, horses and attendant labourers were required if seed drills were to be used, and such costs reduced their attractiveness to many farmers. The machines were also expensive to purchase in the early years and only started to become popular when prices began to drop. To function successfully they required flat, adequately drained land that was clean and well maintained. But even farmers working the relatively flat landscape of the pays de Caux claimed that their land was too undulating for effective use of seed drills (Sion 1909). However, commercial root crops, such as sugar beet, needed to be spaced regularly if they were to grow well, and this requirement provided an essential stimulus for the adoption of drills in northern parts of the Paris Basin. Hence, by 1862 they had become most numerous in the market-orientated ploughlands of the Ile-de-France, Picardy and Flanders, with peak values in Nord (1530) and Pas-de-Calais (1120), where production of sugar beet was flourishing (Fig. 6.6c). Only Finistère and Gard reported more than 200 drills apiece at this time. Thirty years later the same cluster of northern départements contained the largest number of drills (Nord 6185, Pas-de-Calais 5710, Aisne 5480) but important numbers were also encountered in Armorica (Mayenne 3165, Ille-et-Vilaine 2105, Finistère 2890).

Like all other operations, the harvesting of grain was done predominantly by hand throughout the 19th century. Before the 1850s the rural population had provided adequate labour for traditional sickles to be used, with these small implements being manageable by all sections of the workforce, including women, children and old folk (Châtelain 1956). Heavy, high quality scythes had started to be imported during the first quarter of the century but they were expensive and needed strong men if they were to be used efficiently (Tresse 1955). By mid-century they were starting to be adopted, but sickles remained in use for cutting grain in many areas, although scythes might be employed on hay, as was the case in Velay (Merley 1974). In the following decade use of scythes was rapidly replacing use of sickles in the thinly peopled wheatlands of the Paris Basin, but also the era of mechanisation had only just begun to dawn and in 1857 it was prophesied that 'just as the scythe has killed the sickle, so the reaping machine will kill the scythe' (Moreau 1958, p. 159). Nonetheless, reapers were not only expensive but also delicate machines, which had to move across fields that might be rough or hilly, soft or stony, and which needed two or three horses apiece to push them or draw them from the side. For these and other reasons their acceptance proved to be a slow and partial business.

The first statistics were gathered in 1862, when 3050 reapers (*moissonneuses*) were listed but, unlike other machines, major concentrations were in southern France, with départements from Gers to Gard containing two-thirds of the national total (Fig. 6.6d). Six contiguous départements comprised this cluster and hence one may not dismiss this phenomenon as an error in recording. Another focus was in Loire (430), with six other départements in eastern and middle France containing over 20 reapers apiece. By 1892 the pattern had changed drastically and conformed much more closely with the distribution of other arable machines. Fifteen départements from Upper Normandy to Lorraine contained over 500 reaping machines apiece, with highest numbers in Marne (2360), Somme (1170), Oise (1085) and Meuse (1010). These same 15 départements had reported only 4 per cent of the national total in 1862 but by 1892 their share had reached 58 per cent. Elsewhere in France only Mayenne (870), Gers (620) and Charente (505) contained large numbers of reapers in the final decade of the century.

The message of arable mechanisation is intricate but nonetheless clear. Machines cost good money, their iron frames were brittle, moveable parts wore down quickly, and hand-forged replacements seldom fitted well. To be effective and economic many needed to operate over large areas and traverse level fields without boggy patches or large stones. Such conditions were rare, given the structural and environmental contexts of French farming in the late 19th century. Foci of experimentation with various types of machine had displayed a measure of spatial variation during the Second Empire but by 1892 only the plains and plateaux of the Paris Basin and the north-east emerged as significant regions of adoption, with Lower Maine and parts of Aquitaine figuring occasionally. In reality, the application of metallurgy and engineering to agriculture had precious little impact nation-wide and the total number of machines remained remarkably small.

Unfortunately, not all types of machine had been recorded in 1852, but the sum total of the four categories that were listed increased by 183 per cent from 111 825 in 1862 to 317 375 in 1892. Numbers declined in a few southern

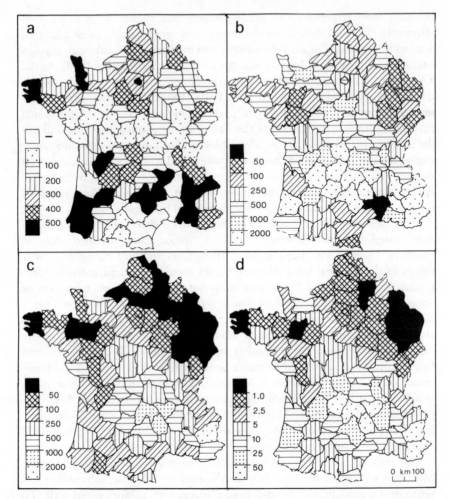

**Figure 6.7**  (a) Change in arable machines, 1862–92 (per cent). Arable land (ha) per arable machine: (b) 1862; (c) 1892. (d) Ratio of farms of more than 10 ha to arable machines, 1892.

départements, possibly because of the resurgence of viticulture and the consequent decline of cereal cultivation (Fig. 6.7a). Low rates of uptake characterised two contrasting types of agricultural environment; some involved districts such as the north-east where mechanisation had occurred quite early and widely; while others, such as the Auvergne, fell at the other end of the farming spectrum, where arable husbandry remained almost totally unmodified. High rates of adoption between 1862 and 1892 were typical of a scatter of départements in the Paris Basin, Armorica, Limousin, the extreme south-west, the southern Massif Central and the lower Rhône, but absolute numbers of arable machines remained small in all except the first two regions.

In 1862 there had been one arable machine to every 229 ha of ploughland in France, with denser concentrations occurring in three broad regions (Fig. 6.7b). Thus in six départements of the north-east there was one machine to less than 100 ha, with a weaker extension in Champagne and Picardy; a second cluster comprised Lower Maine and parts of Armorica; and a feebler third focus appeared in Languedoc. Extremely few machines were found in the Massif Central and the Alps but much more surprising was the strong reliance on manual labour across a wedge of relatively uncomplicated terrain running from Lower Normandy, through Beauce and Perche to middle France (Châtelain 1949). Thirty years later the position had changed both numerically and, to some extent, spatially with there being one arable machine to every 79 ha of ploughland throughout the nation. Mechanisation had become reasonably well established in the north-east, in northern sections of the Paris Basin, Maine and parts of Armorica; however, the early focus in Languedoc had dwindled and disappeared (Fig. 6.7c). By 1892 it would seem that arable machines were having little impact in Lower Normandy or Beauce, and remained virtually absent from the Massif Central, with one machine to over 1000 ha of ploughland in Cantal, Haute-Loire and Lozère.

Mechanisation of arable activities would not have appealed to, nor have been economic on, the small peasant farms that were extremely numerous across many parts of the land. In 1892 the national ratio was one arable machine to every 2.6 farms of 10 ha or more, with above-average mechanisation characterising a group of départements running from the Channel coast, through northern and eastern sections of the Paris Basin to Franche-Comté; and a line of départements extending from Maine into northern Brittany (Fig. 6.7d). The largest and most pronounced focus of arable mechanisation was in Lorraine, where in statistical terms each farm of more than 10 ha could well have possessed at least one arable machine. The reality of the situation was quite possibly otherwise, since mechanised holdings might well have made use of several if not all four types of arable machine, while less progressive farms of comparable size might have had none. At the other extreme, only one farm in more than 50 in the Massif Central and the southern Alps had an arable machine in the final decade of the 19th century (Miège 1961). The long-recognised north–south contrast between progressive and traditional rural areas continued to hold its powerful sway.

# 7 The advance of livestock husbandry

## Changing objectives

The prime objective of agriculture during the ancien régime économique was to produce enough grain to feed an increasing population, but this could only be achieved if adequate manure were available to maintain soil fertility, and so it was to this end that most livestock were raised. In consequence, keeping livestock was not a specialist activity over much of France but rather an essential adjunct to the arable system which, in turn, made available important quantities of feed for sheep through fallowing and stubble grazing. Stretches of moorland and other types of rough land provided pastoral resources which, similarly, were more suited to sheep than cattle. Hence, it is not surprising that Chaptal (1819) estimated that sheep out-numbered cattle by seven to one at the end of the First Empire. Animals of all kinds shared the common function of being dung-making machines throughout their lives; some yielded marketable products, such as milk or wool, on a regular basis; but the full value of others could only be realised when they had been slaughtered. Many livestock changed function through time, for example, spending their working lives as draught animals or providers of milk but ending up as sources of meat or hide. Nonetheless, livestock products did not enter greatly into the human diet in most parts of France in the first half of the 19th century, although patterns of food consumption by Parisians, other townsfolk and by the most comfortable strata of rural society formed exceptions to that rule (Chevalier 1842, De Tocqueville 1843–4, Husson 1856).

Very few districts concentrated on livestock husbandry prior to mid-century and hence relatively little space was devoted to permanent grass, which was a specialised, even 'luxury', form of land use in a rural economy geared to producing the necessary cereals for human subsistence. Well watered vales and humid uplands offered 'natural' advantages for the growth of grass, but until an effective transport system was established to bring in cereals and transport livestock and animal products, the chances of pastoral specialisation remained slight (Dubois 1827). Such conditions were to be transformed after mid-century as reclamation of wasteland, decline of fallows, and advances of fodder crops combined with substantial increases in permanent grassland to produce profound changes in the source, nature and quantity of livestock feed. The agronomic virtues of applying greater quantities of manure to the soil were recognised more widely and consumption of meat and milk by town dwellers rose appreciably, with the emerging railway network enabling such demands to be met from distance regions rather than from districts within the 'urban shadow' as had previously been the case (Houssel 1976). This power of demand came to be expressed most clearly during the final quarter of the century, as market-orientated livestock farming became a more remunerative operation than grain growing. The inequality of reward that might be derived from the two sectors of the farm economy was further accentuated by the forces of change at work during the agricultural depression. Regions of pastoral specialisation developed with a striking clarity, just as emphasis on viti-culture, market gardening or specific aspects of arable cropping came to characterise a growing number of pays by the eve of World War I.

Prefects had not been required to include information on permanent grass in the annual estimates of crop production that commenced in 1815, and hence the first statistics on pastoral resources did not appear until the ancien cadastre and the enquête agricole of the July Monarchy. Permanent grass-land was defined rather differently in each series and considerable divergences may be detected between these roughly contemporaneous sources, largely due to the ambiguous status of poor quality permanent grass. In fact, the cadastral data appear to pose rather fewer problems of interpretation, and hence particular attention will be focused on the evidence that they contain. At the time of the ancien cadastre, 4 612 455 ha were classified as permanent grassland, which covered 9.1 per cent of the country, with important con-centrations in Limousin and Lower Normandy (Vidalenc 1965) (Fig. 2.3a). Indeed, some parts of the latter province had forged important links with the Paris market as early as the 17th century and sizeable areas had been put down to grass by the July Monarchy; Lower Normandy was unquestionably France's leading pastoral focus in the 1830s, with more than a quarter of several arrondissements being under permanent grass (Clout 1980). Very significant increases occurred after 1879, endowing France with no less than 6 664 280 ha of grass in 1907, which covered 13.2 per cent of the total surface. The grassland realm increased by 44.5 per cent over the full span of years from the ancien cadastre to the eve of World War I, with only nine départements in very different environments experiencing net decline (Fig. 7.1a). The reasons for this anomalous trend must have varied considerably between localities and may well have included the spread of market garden-ing or housing across former meadowlands in Seine, Seine-et-Marne and Seine-et-Oise, afforestation in Hautes-Alpes, and the advance of arable cultivation at the expense of ancient grasslands in Limousin.

At the other extreme, permanent grassland increased by large proportions in Bouches-du-Rhône (up 832 per cent) and Seine-Inférieure (up 226 per cent) but still occupied less than one-fifth of those areas in 1908. Eight other départements doubled their grassland realm, although the absolute amount of land under grass in some pays, such as Hérault and Vaucluse, remained

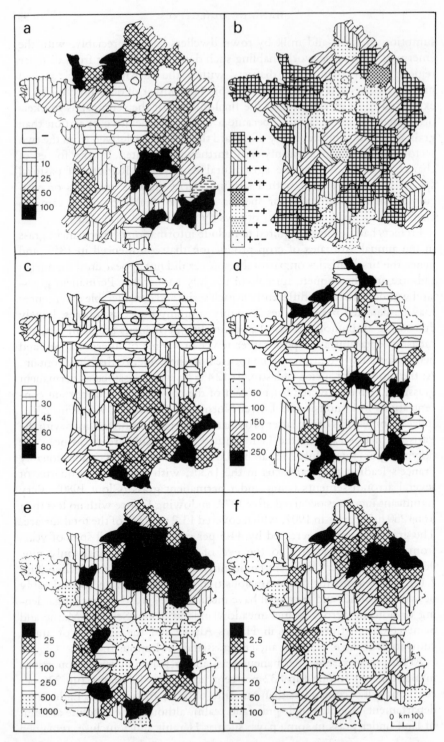

**Figure 7.1** Permanent grass: (a) net change in surface, ancien cadastre–1907 (per cent); (b) typology of change. (c) Area under channel irrigation as percentage of all irrigated meadows, 1892; (d) change in grass output, 1837–1912 (per cent). (e) Ratio of farms of more than 10 ha to grass machines, 1892. (f) Permanent grass (ha) per grass machine, 1892.

very slight. Really significant increases occurred in Manche (up 181 per cent), Cantal (up 190 per cent), Puy-de-Dôme (up 104 per cent) and Loire (up 100 per cent), with 25–45 per cent of each département having been put down to grass by 1907. A very similar pattern emerges when absolute increases are examined, with Manche (up 166 530 ha), Cantal (up 164 020 ha), Puy-de-Dôme (up 108 710 ha) and Orne (up 101 075 ha) heading the league and being followed immediately by Saône-et-Loire (up 98 605 ha) and Calvados (up 80 815 ha). By World War I, grassland farming was concentrated not only in Lower Normandy but also in Auvergne and to a lesser extent in the pays of the Saône and the upper Loire (Boitel 1881a).

## The three phases

The grassland realm retreated fractionally during the July Monarchy but the pattern of change was far from even. In fact, grassland actually advanced in 47 départements but this trend was outweighed by decline in the remaining 35, located mainly in middle France, Champagne and the Paris Basin. Substantial reductions occurred in Haute-Loire (down 830 ha p.a.), Indre (down 575 ha p.a.), Morbihan (down 455 ha p.a.) and Nord (down 295 ha p.a.), with poorish pasture disappearing beneath the plough in three of these départements and wasteland registering an increase in Morbihan (Fig. 7.2a). Positive changes were normally modest during this phase, with only Charente-Inférieure (up 670 ha p.a.) increasing its permanent grass by more than 500 ha each year, being followed by Manche (up 375 ha p.a.) and Basses-Alpes (up 290 ha p.a.) (Fig. 7.2b). Woodland formed the most extensive feature to decline in Charente-Inférieure and that role was performed by wasteland in both Manche and Basses-Alpes. Agricultural practice and rural land use were becoming more orientated to market demands in just a few areas at this time, since the scope for thorough transformation remained slight until the railway network was built.

The nature of change was much more uniform between mid-century and 1879, with the grassland realm advancing in 55 départements and only parts of middle France, the Paris Basin and the far south displaying the reverse trend. Important increases in the pays d'Auge during the 1850s were emulated in Bessin during the 1860s and 1870s and subsequently in many other areas of both Upper and Lower Normandy (Reinhard 1923, Levy-Leboyer 1972). Thus permanent grassland advanced by 1140 ha annually in Calvados between 1851 and 1879 and that focus was ringed by important advances in Manche (up 665 ha p.a.), Orne (up 705 ha p.a.) and Seine-Inférieure (up 345 ha p.a.) (Fig. 7.2c). Arable displayed the most extensive retreat in the first three départements, with woodland undergoing substantial decline in Seine-Inférieure. Districts in southern Armorica, Nivernais, Velay and the Ardennes also formed important foci of grassland

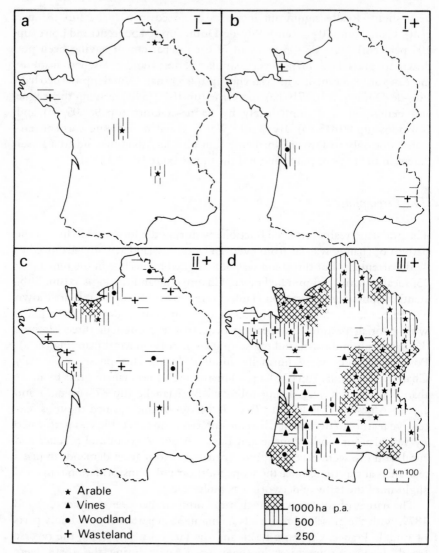

**Figure 7.2**  Permanent grass, annual net change by cadastral phase.

conversion, but in these areas wasteland or woods were the major land–use features that declined (Azambre 1929).

By the two final decades of the century the grassland realm was on the increase in virtually every part of France (Grandeau 1885). It declined in only six départements between 1879 and 1907, and around Paris this may well have been in the face of urban development. More than 1000 ha were being converted to permanent grass each year in no fewer than 18 départements, with peak values involving Cantal (up 5605 ha p.a.), Manche (up 4755 ha p.a.) and Puy-de-Dôme (up 3635 ha p.a.), and extensive changes occurring throughout Auvergne, Burgundy, Franche-Comté and, of course, Normandy (Fig. 7.2d). Some 23 départements increased their grassland by 500–1000 ha annually and a further 15 displayed net increases of 250–500 ha. The Ile-de-France, Champagne and parts of Limousin formed the only significant breaks in this wide expanse of grassland conversion throughout northern, eastern and central France. This transformation occurred at the expense of arable cultivation in almost all parts of the country, although viticulture formed the major element to decline in southwestern France and défrichement of wasteland still continued in a handful of areas.

Only 18 départements fitted the national model of change within the grassland realm, involving an initial period of retreat that was followed by two phases of advance $(-++)$ (Fig. 7.1b). Some 31 départements increased their permanent grass during the whole period $(+++)$, with 11 of these experiencing the fastest rate of grassland conversion after 1879 and the years between 1851 and 1879 coming in second position. These areas formed solid blocks of territory across Lower Normandy, Maine, Nivernais, the Massif Central and the extreme south-west, where commercial livestock husbandry, either for meat or dairying, became increasingly important with the passage of time. In six widely dispersed départements in Brittany, the northeast, Pyrenees and the southern Massif Central the rate of putting down to grass decelerated after 1879, while the third phase was slowest of all in Vendée. Conversion to grass overtook two phases of decline $(--+)$ in 15 départements, being associated with an early development of animal husbandry in the outer Paris Basin and middle France and with fluctuations in viticulture in the Midi. Eleven départements shared in the national advance of grassland after 1879, following a chequered earlier evolution $(+-+)$; and in the remaining 11 départements grassland surfaces actually declined after 1879, in complete contrast with the general trend.

## Means to an end

Relatively little is known about the precise characteristics of grassland in each département, and the somewhat ambiguous distinction between permanent grass and rough grazing is common knowledge. Irrigated meadows were identified in many agricultural enquiries, but not until 1882 was a distinction drawn between those that were irrigated 'naturally' and others that were watered by special channels. Certainly the tradition of channel irrigation was long-established in parts of the relatively arid south and also in more humid areas such as Limousin. The national surface under all kinds of irrigated meadow advanced from 1 460 305 ha in 1852 to 2 355 935 ha in 1892 (up 61 per cent), involving all except three départements in the land. Although large percentage increases occurred in middle France and the Paris Basin, the absolute amounts of watered meadow remained small in these areas. Two-fifths of all grassland was declared to be irrigated in one form or another in 1892, with proportions in excess of 60 per cent characterising Limousin, Nivernais, the Vosges, Vaucluse and Lower Brittany. By contrast, less than 20 per cent of the grassland in Normandy and the northern Paris Basin was subject to irrigation.

Meadow irrigation was most extensive in humid areas such as Armorica, the Massif Central, Pyrenees and Vosges, with fewer than three dozen départements containing over 25 000 ha of irrigated grassland apiece in 1892 (Boitel 1881b). The figure exceeded 50 000 ha in 11 areas in the Massif Central and the lower Loire, to reach a peak of 91 000 ha in Saône-et-Loire. The importance of the Massif Central emerges overwhelmingly when attention is focused specifically on channel irrigation, with Haute-Vienne (61 260 ha) and Cantal (51 870 ha) coming at the head of the league, and every département in the region having above-average (44 per cent) proportions of their irrigated meadows watered by special channels (Fig. 7.1c). Channel irrigation proved particularly essential in four southern départements (Hautes-Alpes, Bouches-du-Rhône, Vaucluse, Pyrénées-Orientales, each exceeding 80 per cent), although the amounts of land involved were slight. As a general process, however, meadow irrigation displayed a very different pattern because of the overwhelming importance of riverside flooding in northern and middle France.

Grassland yields varied greatly between districts and from season to season but the enlargement of the grassland realm produced a very substantial increase in fodder resources, with the annual output of grass rising by 127 per cent from 100 952 700 qx in the late 1830s to 229 933 700 qx in 1912. Three-quarters of the latter figure derived from hay meadows and that source accounted for virtually all the grass that was produced in 26 départements in the Ile-de-France and to the south of the Loire. At the other extreme grazed pastures (*herbages*) yielded more than half of the grass output in Normandy and the northern Paris Basin, surpassing 75 per cent in Calvados, Nord, Pas-de-Calais and Seine-Inférieure; while the pastoral region of Nivernais derived roughly one-third of all its grass from herbages. With the exception of three départements at the heart of the Paris Basin, every part of France increased its grass output from the 1830s to 1912, with yields rising by more than 250 per cent in Normandy, Nord, the lower Rhône, the middle Garonne and the northern Massif Central, and above-average increases accruing throughout northwestern France, Nivernais, and Charolais (Fig. 7.1d). On the eve of World War I, hay meadows and herbages in Normandy, Picardy, the lower Rhône and parts of the south-west were yielding more than 50 qx/ha each year but in terms of absolute output the dominance of Normandy was overwhelming. Four of the six départements that generated more than 6 000 000 qx each year were located in that province, which accounted for no less than one-sixth of the national total. Calvados (12 896 000 qx), Manche (7 370 000 qx), Orne (6 049 000 qx) and Seine-Inférieure (6 034 000 qx) headed the league, with a grand total of 31 départements in the north-west, Massif Central, Franche-Comté and the extreme north producing more than 3 000 000 qx apiece each year.

Hay making was overwhelmingly accomplished by hand during the second half of the 19th century but reapers and tedders were employed on a minority of holdings. Both were rare in 1862 with a mere 4720 reapers and 2655 tedders being recorded throughout the land. However, 15 départements contained more than 100 reapers apiece and these were mainly located in Languedoc, with extensions in Aquitaine and the valleys of the Rhône and the Saône, and outliers in Calvados and Pas-de-Calais. By contrast, only 31 were recorded in Seine-et-Marne and Seine-et-Oise combined. Just five départements contained more than 100 tedders apiece, namely Tarn (115), Saône-et-Loire (175), Calvados (310), Pas-de-Calais (310) and Aude (865), and the resultant pattern suggests early foci of adoption in Upper Languedoc and in the markedly pastoral environments of Lower Normandy, Boulonnais and the Saône valley. The early distribution of both of these types of implement was quite different from that of arable machines in the early 1860s. By 1892 substantial changes had occurred, with 11 départements from Upper Normandy to Lorraine each containing more than 1000 reapers (reaching a peak of 3115 in Marne), but beyond the Paris Basin only Mayenne (1900) and three départements in Aquitaine surpassed the 1000 mark. Tedders had become numerous throughout the Paris Basin, with peaks in Seine-Inférieure (2975), Meuse (3170) and Marne (2905), and more isolated foci of adoption in the Charentes and Upper Languedoc. In spite of very different origins, subsequent processes of adoption have rendered the patterns of mechanised hay making and arable husbandry strikingly similar at the end of the 19th century.

The 90 040 hay-making machines recorded in 1892 represented an average density of one machine to every 66 ha of permanent grass. Enormous contrasts were concealed by that figure, ranging from less than 12 ha per machine in Champagne (Marne 6 ha, Meuse 11 ha) and the Ile-de-France (Seine-et-Marne 9 ha, Seine-et-Oise 9 ha) to over 1000 ha in Brittany, much of the Massif Central and Hautes-Alpes, where of course most permanent grass was used for grazing rather than hay cutting (Fig. 7.1e). The same point also applied to the herbages of Lower Normandy and partly explains the relatively low level of mechanisation encountered there. On average, there was one tedder or reaper to every 9.2 farm holdings of more than 10 ha in 1892. As one might expect, hay making on 'large' farms in the eastern Paris Basin and Upper Normandy displayed important aspects of mechanisation, with roughly one machine to every other farm (Meuse 1.8, Marne 2.1, Seine-Inférieure 2.4, Seine-et-Marne 2.4, Meurthe 2.5) (Fig. 7.1f). North-eastern France as a whole recorded one machine to every five 'large' farms but similar ratios were only encountered elsewhere in Charente, Charente-Inférieure and Mayenne. At the other extreme, less than one farm of more than 10 ha in every hundred had any kind of hay-making machine in western Brittany or parts of the Massif Central.

Artificial meadows have rightly been considered as a component of the arable realm but now it is appropriate to examine the fodder that they yielded in the context of pastoral husbandry. During the 1830s the combined yield of clover, lucerne and sainfoin amounted to a modest 45 812 900 qx p.a. but by 1912 it had risen by 166 per cent to 122 081 000 qx even though output had decreased in Nord and three other départements. Rates of increase proved to be below average in the Paris Basin, where artificial meadows had been accepted reasonably widely during the 1830s, but rates were very high in northwestern, central and southern France, where artificial meadows had been much less known at that time. On the eve of World War I, artificial meadows in 25 départements were each producing more than 2 000 000 qx of fodder each year, with a cluster of such areas covering much of the Paris Basin, but with peak production occurring in Vienne (4 090 000 qx), Seine-et-Marne (3 160 000 qx), Charente-Inférieure (3 010 000 qx) and Isère (3 000 000 qx). Fodder output from permanent grass and artificial meadows rose substantially and many other fodder crops were being widely grown by the early 1900s but unfortunately it is not possible to trace their development over the preceding three-quarters of a century. In spite of such important advances, the evolution of fodder production was by no means all on the positive side. Clearance of wasteland and abandonment of fallowing in many districts removed very important traditional sources of feed, and this fact had profound implications not only for the types of livestock that might be reared but also for the general distribution of animal husbandry.

## Changing numbers

France had supported 31 550 010 sheep in the late 1830s and the total rose to 32 657 375 in 1852 but decreased thereafter to 16 014 940 in 1912, which represented a fall of 49 per cent in 75 years (Tables 7.1 & 2). No fewer than 44 départements had recorded their peak values numbers in 1837, with a further 33 maxima occurring in 1852 (Block 1850) (Fig. 7.3a). The story of declining

**Table 7.1**  National livestock totals.

|            | 1837       | 1852       | 1862       | 1882       | 1892       | 1902       | 1912       |
|------------|------------|------------|------------|------------|------------|------------|------------|
| sheep      | 31 550 010 | 32 657 375 | 28 680 920 | 23 086 790 | 20 489 200 | 17 924 810 | 16 014 940 |
| pigs       | 4 593 975  | 4 969 720  | 5 650 270  | 6 978 250  | 7 236 640  | 7 024 240  | 6 753 250  |
| horses     | 2 663 635  | 2 710 470  | 2 759 360  | 2 800 575  | 2 762 575  | 2 992 195  | 3 186 750  |
| all cattle | 9 533 350  | 13 536 020 | 12 046 125 | 12 623 440 | 13 338 590 | 14 560 365 | 14 343 430 |
| cows       | 5 252 835  | 5 516 565  | 5 970 420  | 6 370 660  | 6 470 865  | 8 088 360  | 7 520 710  |

**Table 7.2** Peak livestock numbers (départements).

|           | 1837 | 1852 | 1862 | 1882 | 1892 | 1902 | 1912 |
|-----------|------|------|------|------|------|------|------|
| sheep     | 44   | 33   | 3    | 1    | 1    | 0    | 0    |
| pigs      | 1    | 3    | 7    | 24   | 16   | 15   | 16   |
| horses    | 10   | 5    | 4    | 3    | 2    | 6    | 52   |
| all cattle| 4    | 0    | 5    | 3    | 3    | 16   | 51   |
| cows      | 3    | 1    | 9    | 0    | 0    | 56   | 13   |

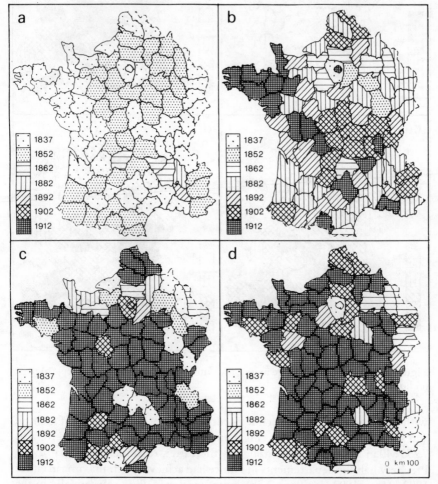

**Figure 7.3** Dates of peak livestock numbers: (a) sheep; (b) pigs; (c) horses; (d) cattle.

sheep numbers was similar throughout France, although there were local variations in intensity and chronology (Zolla 1885). Vigorous reductions were associated with the retreat of bare fallow in the northern half of France,

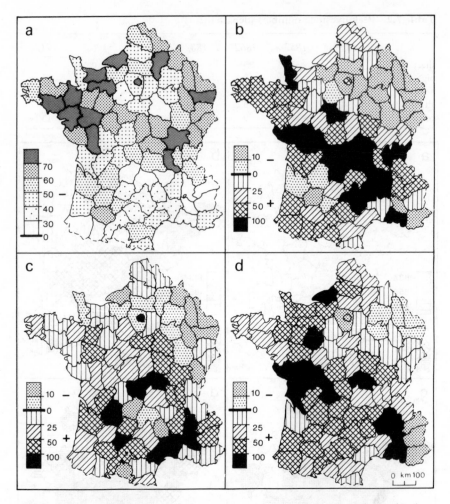

**Figure 7.4** Percentage change in livestock numbers, 1837–1912: (a) sheep; (b) pigs; (c) horses; (d) cattle.

with more gentle rates of decline occurring in the Alps, the Massif Central and the Pyrenees (Fig. 7.4a). Pig numbers rose from 4 593 975 in 1837 to 7 236 640 in 1892 and retreated slightly thereafter, reaching 6 753 250 in 1912. Precisely half the départements in France had achieved their maxima in either 1882 (24) or 1892 (16) (Fig. 7.3b), with numbers falling substantially in the north-east and the Ile-de-France (Fig. 7.4b). By contrast, the pig population of the north-west, much of southern France and the Massif Central continued to increase markedly.

France had contained 2 663 635 horses in the late 1830s and this total had risen by one-fifth to 3 186 750 in 1912, with 52 départements recording their peak values on the eve of World War I (Fig. 7.3c). However, many areas in the northern third of the country, and especially in the north-east, contained fewer horses in 1912 than 75 years previously, possibly because of farm amalgamation and the retreat of arable with subsequent increases in the amount of land under permanent grass or trees (Fig. 7.4c). Exactly the reverse trend occurred in départements to the south of a line from Normandy to Lake Geneva, in which triennial rotations, cultivation of oats and fodder crops, and the use of horse-drawn ploughs had advanced in concert during the 19th century (Anon. 1867). However, the really crucial change in livestock husbandry was the rise in cattle numbers, with cattle of all kinds increasing from 9 533 350 in the late 1830s to 14 560 365 in 1902, being followed by a slight reduction to 14 343 430 in 1912. No fewer than 67 départements achieved their peak numbers in 1902 (16) or 1912 (51), but maxima were recorded as early as 1862 in five northeastern départements in response to the profound changes in land use and agricultural activity which affected that region (Fig. 7.3d). Cattle numbers also contracted in the Alps but rose very substantially in Normandy, along the fringes of Armorica, and in the Massif Central, Poitou and the south-west with totals increasing by more than 100 per cent in a dozen départements (Fig. 7.4d).

These intricate changes may be summarised by computing indices which embrace all types of livestock and by weighting their significance in the total according to accepted coefficients. The technique that was devised by Joseph Klatzmann (1955) to synthesise numbers of horses, pigs and sheep as well as cattle into *unités bovins* will be extended to incorporate asses, goats and mules and thereby determine 'livestock units' (lu).[1] In the late 1830s the French agricultural system supported 18 494 260 lu at an average density of 37.0/km² of total surface, by 1912 the figure had risen by 24 per cent to 22 917 060 giving a density of 45.8/km². At the start of the period, 36 départements displayed densities of 30–39 lu/km², which fell close to the national average, even though very different rural mosaics in regions as diverse as Champagne, middle France, the Massif Central and Aquitaine were involved (Fig. 7.5a). Stocking rates were above average in a crescent of départements from Armorica, through Normandy and the northern Paris Basin to northeastern France, and this pattern evokes certain similarities with the map of human densities at that time (Fig. 2.1d). There were, however, important differences in detail, with very high concentrations of livestock not only occurring in Seine (159/km²) as might be expected, but also in Finistère (71/km²), Nord (69/km²), Manche (68/km²) and Côtes-du-Nord (66/km²). At the other extreme, the limited grazing resources of eight southern départements in the Landes, Languedoc and Provence supported fewer than 20 lu/km².

Goats, mules and asses together comprised only 3.7 per cent of national livestock units in the 1830s, and their collective role was to decline substantially by the early 20th century. The really important transformation that

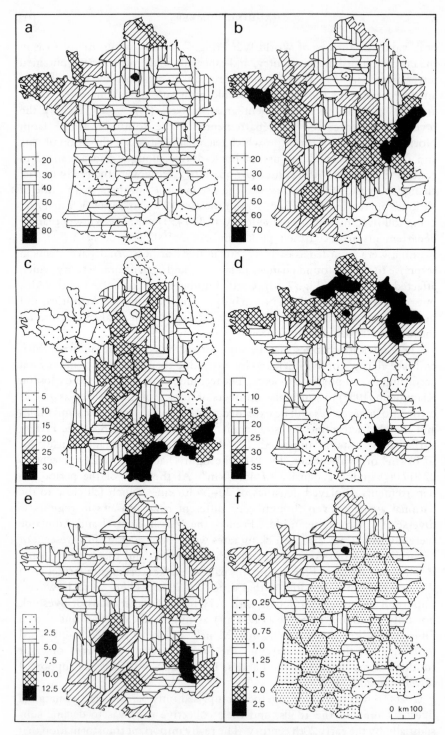

**Figure 7.5** (a) Density of all livestock units, 1837 (lu/km²). (b) Cattle, (c) sheep, (d) horses and (e) pigs as percentage of livestock units, 1837. (f) Livestock location quotients, 1837.

occurred between the July Monarchy and the eve of World War I comprised the advance of cattle farming and the retreat of sheep husbandry. Calculation of livestock units, of course, consistently scales down the importance of sheep and inflates that of cattle in a way that is roughly commensurate with their body weights and thus presents a more dramatic picture than would be derived from considering absolute figures in their untransformed state. Cattle made up roughly half (51.5 per cent) of the national livestock units total during the 1830s and performed a variety of roles including being draught animals as well as providers of milk, meat and hides. Cattle comprised 60–70 per cent of livestock units throughout the uplands of Auvergne and Nivernais and the humid countrysides of Armorica, passed 70 per cent in five départements and reached twin peaks in Ain and Jura (both 78 per cent) (Fig. 7.5b). The presence of considerable numbers of horses and sheep served to reduce the significance of cattle in the Paris Basin and large numbers of sheep in middle France and the Midi produced exactly the same effect.

With a very low coefficient for calculation, sheep comprised only 17.0 per cent of livestock units during the 1830s and were poorly represented in cattle country (Fig. 7.5c). By contrast, they made up more than a quarter of the total livestock units in Beauce, Brie, middle France and much of the south, surpassing two-fifths of the total in Lozère (41 per cent), Pyrénées-Orientales (42 per cent), Basses-Alpes (43 per cent), Aude (43 per cent) and Bouches-du-Rhône (44 per cent) to reach 52 per cent in Hérault. The method of calculation tended to emphasise the importance of horses, which made up 21.6 per cent of all livestock units. They were few and far between in central France, where mules served as beasts of burden and oxen were often used for ploughing, but were particularly numerous in northern départements, arching from Finistère through Normandy and Picardy into the north-east (Fig. 7.5d). Horses represented more than 40 per cent of livestock units in Meurthe and Meuse, reaching a surprising 63 per cent in Gard and no less than 72 per cent in Seine, which was far less unexpected. Finally, pigs accounted for a mere 6.2 per cent of livestock units but were particularly significant in Lorraine, the Rhône valley and a number of pays fringing the Massif Central (Fig. 7.5e). Only in Drôme (14 per cent), Dordogne (15 per cent) and Vaucluse (16 per cent) did they exceed double the national average.

Calculation of location quotients provides a convenient way of comparing the actual distribution of livestock units with what might have been expected simply from considering the total area contained in each département.[2] The northern pastoral crescent emerges remarkably sharply from Figure 7.5f, with Armorica, Upper Normandy, Artois, Flanders and the immediate environs of Paris each supporting livestock densities that were at least 50 per cent greater than might have been expected (Dubost 1876). Franche-Comté and parts of the Massif Central also supported higher than expected densities (Lebeau 1955). At the other extreme, stocking rates were low throughout central and southern France, and fell to below half the expected level in the

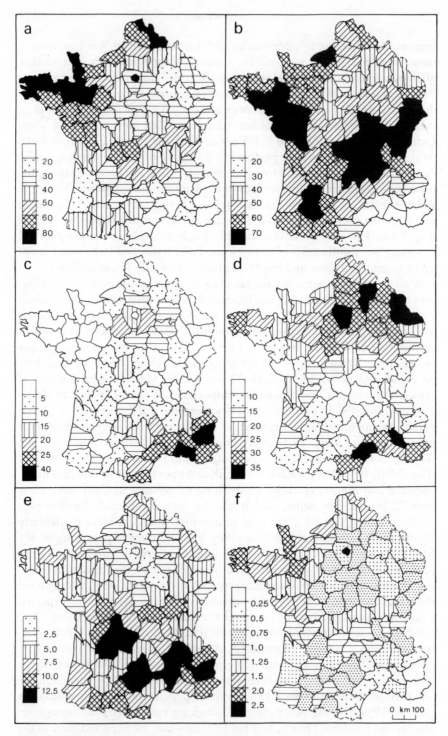

**Figure 7.6** (a) Density of all livestock units, 1912 (lu/km²). (b) Cattle, (c) sheep, (d) horses and (e) pigs as percentage of all livestock units, 1912. (f) Livestock location quotients, 1912.

southern Alps and Hérault. An important contrast existed within the Paris Basin, with livestock numbers in northern and western areas exceeding the expected levels but falling below them to the south and east of the capital.

Three-quarters of a century later, not only had the national total of live-stock units increased by one-quarter but also important changes had taken place in the relative significance of individual types. Only Seine had supported more than 80 lu/km² in the 1830s, but by the eve of World War I no less than seven départements surpassed that figure, with Seine (321 lu/km²) still at the top of the league and being followed by Manche (99 lu/km²), Finistère (98 lu/km²), Mayenne (90 lu/km²), Ille-et-Vilaine (85 lu/km²), Côtes-du-Nord (83 lu/km²) and Nord (81 lu/km²) (Fig. 7.6a). A further 11 départements supported 60–80 lu/km², to encompass the whole of Armorica, Artois, Limousin and much of Normandy. The pastoral role of the humid west was confirmed much more strongly than before. Cattle represented 62.6 per cent of all livestock units in 1912 (Table 7.2) and exceeded 70 per cent over one-quarter of the country, which included Armorica, Auvergne, Franche-Comté and Gascony, with 80 per cent being surpassed in Ain (81 per cent), Vendée (82 per cent) and Jura (83 per cent) (Fig. 7.6b). The basic outlines of French cattle country had been in evidence during the July Monarchy but became much more clear and intense in subsequent decades.

As fallows contracted in the Paris Basin and wasteland was reclaimed across middle France, so sheep rearing declined drastically from 17.0 per cent of national livestock units in the 1830s to 7.0 per cent in 1912. This involved a contraction of sheep-rearing country to its bastions in Languedoc and Provence, where sheep still accounted for more than one-quarter of livestock units on the eve of World War I and even reached 41 per cent in Basses-Alpes and Bouches-du-Rhône (Fig. 7.6c). The significance of horses in the national total of livestock units declined fractionally from 21.6 per cent to 20.9 per cent over the same period. Although overall change was slight, there were significant regional modifications, with a more pronounced stretch of horse husbandry appearing in the Mediterranean south (Fig. 7.6d). By contrast, the importance of horses declined in Brittany and Normandy as cattle rearing advanced in both regions and the arable realm retreated sharply in Lower Normandy. Finally, pigs had increased fractionally in importance to account for 7.4 per cent of total livestock units in 1912, with major concentrations in Périgord, Limousin and the southern fringe of the Massif Central, and more than twice the national proportion being surpassed in Aveyron, Basses-Alpes (both 15 per cent), Corrèze and Drôme (both 16 per cent) (Fig. 7.6e).

The emphatic rise of cattle husbandry in Armorica and to a lesser extent in the Massif Central combined with the contraction of sheep rearing in all but the extreme south to change profoundly the pattern of French livestock farming between the 1830s and the eve of World War I. In 1912 four départements supported more than double the volume of livestock units that considerations of area alone might suggest (Fig. 7.6f). The vast demand for meat and dairy products in the capital produced a truly exceptional location quotient of 7.4 in Seine, which was followed by exceedingly important values in Finistère (2.2), Manche (2.2) and Mayenne (2.0). The crescent of livestock country that had stretched across northern France in the 1830s would appear to change its configuration during the subsequent three-quarters of a century, by shifting away from the départements of the north-east (all of which recorded location quotients below unity in 1912) and developing more emphatically in Armorica, Normandy, Poitou and the Massif Central.

## Notes

[1] In order to calculate *unités bovins*, Joseph Klatzmann multiplied each head of livestock by the following values: all cattle (× 1.0), sheep (× 0.1), pigs (× 0.25), horses (× 1.5). These multipliers have been used in the present study, to which data have been added on goats (× 0.1), asses (× 0.75) and mules (× 0.75) to calculate 'livestock units'.

[2] A location quotient below unity indicates a relative absence of that population in the département, while values above unity indicate greater than expected proportions.

# 8 Polyculture to specialisation

## More a way of life

Viticulture had known a long and honourable past in pre-Revolutionary France, with lord and peasant utilising their expertise to produce the fruit of the vine in a wide range of environments, including many that would now be judged too marginal for successful cultivation (Dion 1959). The sale of wine provided a welcome source of income in the local economy, both for great estates that dispatched wine to distant cities or overseas (and often relied on water transport in the absence of adequate roads), and for small peasant farms that sold surpluses to their nearest market town or exchanged them for grains or other foodstuffs from neighbouring pays. Viticulture produced a distinctive *genre de vie* in many parts of the country and allowed relatively high densities of population to survive in areas where alternative resources were few and far between. Vineyards required extra inputs of labour during the *vendange*, and migrant harvesters, often from harsher districts, were frequently employed. For example, bands of mountain folk descended from the highlands of Tarn to labour in the vineyards of Languedoc; a share of their wages certainly returned home with them, although some commentators complained that too much was squandered on alcohol (Inspecteurs: Tarn 1845). Steep slopes could be terraced and poor soils enhanced to support the vine, thereby rendering productive environments that were quite unsuited to growing cereals. The demand for wine had risen substantially in the final decades of the ancien régime and so much land was being converted to vines that many feared that grain growing would suffer. Further extension of viticulture had in fact been prohibited by law in 1731 but this had little effect, with the freedom to cultivate land as one wished, which was proclaimed at the time of the Revolution, providing a powerful stimulus for even more land to be put under vines.

According to some contemporaries, viticulture was taking land that was needed for grain; for others, it was lamentable that vines were being planted where insolation and soil were inadequate to yield palatable wine. According to Lullin de Châteauvieux (1843), 'stony soils, scrub and rock have been converted into vineyards and thus it continues virtually every day. No concern is shown for the quality of the wine to be produced; the only objective is to increase the quantity indefinitely' (p. 444). Most modes of transport remained poor during the first half of the 19th century, but the vineyards of Bordelais and Burgundy were linked to the capital by reasonably effective water routes and it was these areas that were able to respond to growing Parisian demand for *vin ordinaire* (Vidalenc 1970). By contrast, pays located further south only started to realise their viticultural potential during the railway age. Prior to mid-century, vine-growing had been just one element in local polyculture in Beaujolais, Languedoc and many other districts but subsequent improvements in communication gradually enabled farmers in these areas to grow vines in increasingly sure knowledge that the wines they produced would reach their destinations in a drinkable condition (Latreille 1975, Sorre 1907). Rising numbers of urban dwellers generated a formidable increase in demand as the 19th century wore on, and this trend was further accentuated by a significant change in behaviour which made wine an acceptable drink for women and children as well as grown men. Yet *vignerons* could never be sure of success. Vagaries of weather could provoke great fluctuations in yield, with years of shortage ruining those with no wine to sell and enriching those with surpluses of which to dispose. Fluctuations in quantity affected different qualities of wine in different ways; hence markets could almost certainly be found for fine wines but such was not the case for ordinary wines during phases of glut. In the second half of the century, a range of diseases brought widespread disaster to many viticultural districts and triggered off a variety of responses, which ranged from abandonment and despair to costly reconstitution. The vigneron who marketed his wine had to be something of a gambler, with viticulture most certainly functioning as the joker in the pack of national land uses.

Some 1 520 600 ha were estimated to have supported vines on the eve of the Revolution and planting continued thereafter at an average rate of 11 740 ha p.a. for almost 50 years to raise the national total by one-third (Block 1860). As a result, 2 073 870 ha of vineyards were registered in the ancien cadastre and 1 937 870 ha in the agricultural enquiry of the late 1830s. In fact, the process of change had been far from consistent during the preceding half century, since viticulture had retreated in no less than 15 départements which were close to the northern limit of cultivation or were located in the Alps or the Massif Central (Avannes 1834, Musset 1908) (Fig. 8.1a). The harsh winter of 1789 had killed numerous vines, which were duly uprooted in many marginal areas, and precisely the same result followed other severe winters (Salmon 1840–1). Vines that did survive in such localities often produced very poor results, such as the wines of Caen which 'were diabolical to taste and brought tears to the eyes, being worse than vinegar' (Jardin & Tudesq 1972, p. 22).

By contrast, between 1788 and the late 1830s, vine growing advanced in Burgundy, parts of the Ile-de-France and the north-east and virtually all of southern France, with nine départements augmenting their vineyards by more than 65 per cent and increases of more than 50 per cent being achieved in Côte-d'Or, Charente-Inférieure, Dordogne, the pays of the middle Garonne and parts of the Massif Central. Maximum absolute increases were recorded in Charente-Inférieure (up 42 845 ha) and Hérault (up 41 575 ha), with fresh planting in the top eight départements accounting for no less than

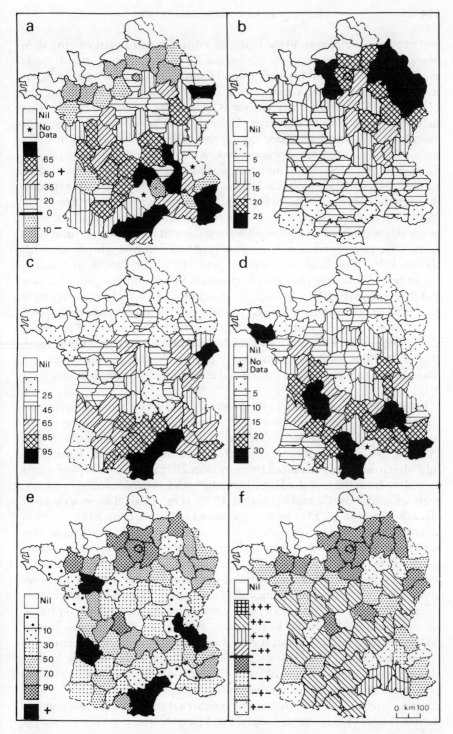

**Figure 8.1** Vines: (a) percentage net change in surface 1788–ancien cadastre; (b) number of vines/ha, 1852; (c) percentage of reconstitution achieved with American vines, 1882–92; (d) newly-planted vines as percentage of total, 1892; (e) percentage net change in surface, ancien cadastre–1907; (f) typology of change.

half of all new vineyards. In the Charentes this reflected a marked growth in overseas demand for *eau de vie* during the Restoration period, which made the fortune of many landowners around Cognac and Angoulême, and contrasted starkly with the disruption of trade that had occurred during the Napoleonic Wars when many vineyards had been uprooted and the land that they occupied put down to cereals (Le Play 1879). Likewise, the advance of viticulture in Lower Languedoc during the 1830s heralded a trend that was soon to become even more pronounced, with Aramon vines being planted for their high yields and Carignan strains being used for their high alcohol content (Marres 1935, Wolff 1967). As a result, the volume of wine exported through the port of Cette in 1844 was precisely twice the quantity handled in 1818. Vineyards also replaced olive groves and other forms of land use in the lower Rhône, but while vignerons were willing to lavish care on their vines when wine prices were high, they displayed less attention when returns fell (Nicod 1956). This kind of behaviour, together with variations in weather, might help explain the profound fluctuations in wine output from that area and others (George 1935). In 1829, no fewer than 2 079 000 landowners were estimated to possess vines, which, according to the ancien cadastre covered 4.1 per cent of France, with very high proportions in Hérault (16.8 per cent), Charente (17.1 per cent) and Charente-Inférieure (18.1 per cent) (Block 1860) (Fig. 2.3b). But, of course, vines were planted far less densely in the arid Midi than in moister areas of central and especially northeastern France (Fig. 8.1b).

In 1851, France contained 2 142 810 ha of vines, with vineyards having increased in 42 départements during the preceding 15 years. However, the national rate of increase had decelerated to 4310 ha p.a. by comparison with 11 740 ha p.a. in the half century that followed the Revolution. During the July Monarchy, vines advanced by more than 250 ha p.a. in six départements extending from the mouth of the Rhône through Languedoc to the Pyrenees, with an outlier in Gironde (Laurent 1978) (Fig. 8.2a). The vineyards of Hérault advanced by 1815 ha p.a. (far ahead of Gironde, up 595 ha p.a.) as a veritable revolution took place in the rural economy of Lower Languedoc, with the invading flood of vines replacing cereal crops and olive trees across much of the plain (Sorre 1907). The environs of Narbonne and other localities which had generated surpluses of cereals in the earlier years were given over increasingly to vines as grain prices dipped at mid-century (Sentou 1947–8). At a finer scale, vine growing extended up slope in parts of Lyonnais and the Monts du Beaujolais to conquer sunny patches as high as 600 m (Garrier 1973). Cadastral data do not allow the precise mechanisms of land-use change to be known, but it is clear that arable was the leading element to decline in Hérault and Pyrénées-Orientales, and wasteland performed that role in many other southern départements at this time. By contrast, virtually all vineyards in the northern third of France continued to decline (Legoyt 1863). Aube, Côte-d'Or, Haute-Marne and Yonne were the

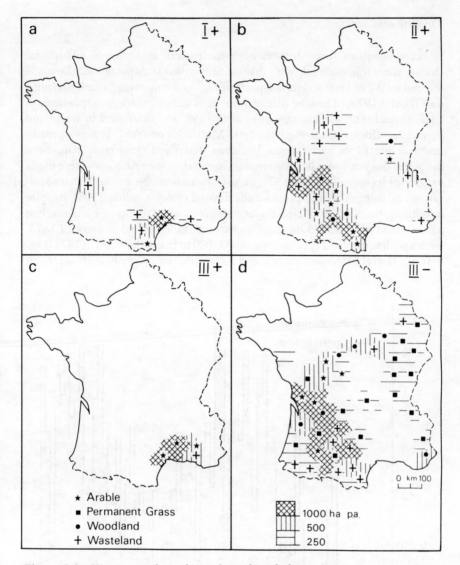

**Figure 8.2**  Vines, annual net change by cadastral phase.

only real exceptions but their collective advance amounted to a mere 2600 ha in one and a half decades, which was less than two years' average net increase in Hérault! Major losses were recorded in Loiret (down 4292 ha), Saône-et-Loire (down 3000 ha), Seine-et-Oise (down 4635 ha) and especially Var (down 23 460 ha).

## Decline and fall?

In the subsequent three decades both the pattern and pace of viticultural change were transformed, with disease, improved transportation, the commercial treaty of 1860 and new aspects of regional competition each playing a part (Barral 1979). The vine disease known as oïdium made its appearance in 1848, spread with increasing vigour after 1850, and flourished in warm and humid conditions in some parts of the Midi (Guyot 1868). It was virtually undetectable in its early stages but then destroyed the grape crop, even though it did not cause permanent damage to the vine. Attempts were made to control its spread and, in 1856, Henri Mares was able to show that applications of sulphur would produce the desired effect (Guillon 1905). In spite of oïdium, the amount of land under vines continued to increase and reached a maximum of 2 538 495 ha according to the international enquiry of 1873, with a peak wine yield approaching 80 000 000 hl being recorded in 1875 (Fig. 8.3). Cadastral land–use figures failed to capture such dizzy heights since, by

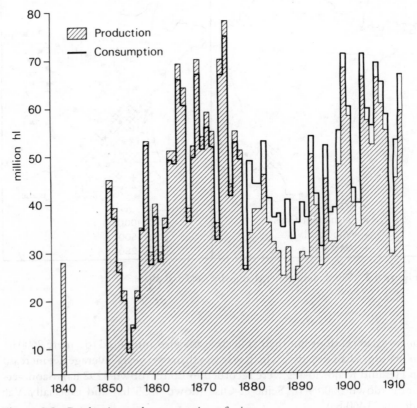

**Figure 8.3** Production and consumption of wine.

the time of the 1879 revision (2 282 300 ha), many viticultural districts were experiencing the ravages of a much more devastating enemy.

Phylloxera had begun to make its impact felt at Pujault (Gard) in 1863, affected the lower Rhône in 1864 and then spread throughout the south-east, so that by 1870 vine growing was on the decline in six southern départements (Stevenson 1980). Vineyards in other parts of France were afflicted at later dates, with those in Aisne, Aube and Haute-Marne not being infected until the 1890s (Colin 1973). Nonetheless, one-quarter of the nation's vineyards had been contaminated by 1878, when the government legislated to establish *cordons sanitaires* around infected districts and to prohibit transport of vines through them (Barral 1979). Southern départements that had undergone massive viticultural expansion earlier in the century were now losing extensive stretches of diseased vines, which were being uprooted by their owners (Huetz de Lemps 1978). No less than 70 835 ha were lost in Gard between 1851 and 1879 (down 2530 ha p.a.), while net losses amounted to 51 000 ha in Hérault (down 1825 ha p.a.) and 45 815 ha in Bouches-du-Rhône (down 1635 ha p.a.). These net figures undoubtedly provide an under-statement of the true volume of decline, since new planting continued in each of these districts between 1851 and the early 1860s and hence the real volume of loss must have been substantially greater. A separate patch of phylloxera had developed near Bordeaux in the early 1860s, but this did not coalesce with the major zone of infection until 1880 (Guillon 1905).

Many other parts of the viticultural realm managed to escape the devastation of phylloxera during the second cadastral phase. Between 1851 and 1879, surfaces under the vine continued to expand in many areas, with a dozen départements along the Loire and in central-western France experiencing a striking change in fortune following a phase of retreat during the July Monarchy. No fewer than 19 départements managed to increase their vineyards by more than 250 ha p.a. between 1851 and 1879 and very impressive advances took place to the west of the main phylloxerated zone in Aude (up 2320 ha p.a.), Haute-Garonne (up 5655 ha p.a.), Gironde (up 1205 ha p.a.) and Dordogne (up 1075 ha p.a.) (Fig. 8.2b). Construction of the Bordeaux–Cette railway during the second half of the 1850s had improved communications greatly in the first three départements and this, complemented by the early impact of phylloxera in eastern Languedoc, provided a powerful stimulus for new planting (Sentou 1947–8). Olive groves, arable fields and rough grazing were displaced by vines in the Corbières and a real 'fury' of planting spread over the land between the Tarn and the Garonne (Brunet 1954, 1959). This territory was to be spared the ravages of phylloxera until the 1880s, by which time vine growers were able to take advantage of technical expertise developed elsewhere to combat the disease. The yield from many new vineyards proved to be abundant but of low quality, as in the environs of Toulouse where many cultivators appeared to lack adequate skill or interest to produce decent wine (Jorré 1971). Other land-use realms retreated substantially in many southern départements as extensive areas of vines were planted. Ploughland was the major loser in Aude and Dordogne, with wasteland suffering that fate in Gironde, and woodland in Haute-Garonne.

After 1879 the fortunes of the vine changed once again as phylloxera reached all viticultural areas and dramatic reconstitution was undertaken in the Midi méditerranéen. In 1882, vineyards in 53 départements had been affected, two years later the total was 67, and by 1892 phylloxera had reached as far north as Epernay (Guillon 1905). The devastation that followed in its train was nothing short of a national disaster, with perhaps 400–500 000 ha of vines being destroyed (Laurent 1976). In addition, black rot caused a new wave of devastation after 1885. Wine output plunged to just over 20 000 000 hl in the late 1880s, the value of phylloxerated vineyards plummetted, and whole pays were exposed to ruin (Fig. 8.3). The national total of land under the vine slumped by one-third, from the peak of the early 1870s to 1 752 120 ha in 1892, 1 726 050 ha in 1902, 1 479 035 ha in the cadastral revision of 1907, and 1 590 695 ha in the agricultural enquiry of 1912. Retreat was virtually a nationwide phenomenon and only five départements managed to increase their vineyards between 1879 and 1907. These were precisely the areas where phylloxera had struck in the 1860s and a considerable measure of re-adjustment, reconstitution and recovery had subsequently come into play. Between 1879 and 1907, Gard and Hérault increased their vineyards by 63 715 ha and 92 385 ha respectively, being followed by Bouches-du-Rhône (up 18 735 ha), Vaucluse (up 21 450 ha) and Drôme (up 3960 ha). Arable land declined markedly in most of these départements to compensate for this advance, although wasteland was the leading element to retreat in Bouches-du-Rhône (Fig. 8.2c).

In complete contrast, viticulture declined massively in the south-west where départements lost more than 1000 ha each year, which represented a combined loss of 497 910 ha between 1879 and 1907 (Fig. 8.2d). Vine grow-ing had occupied 17.1 per cent of Charente in the ancien cadastre but a mere 4.0 per cent in 1907 (Bernard 1978). A whole way of life was over! Arable cultivation made an important advance over former vineyards in that département and in neighbouring Charente-Inférieure, but in the Garonne valley permanent grassland displayed the most significant increases, and woodland performed that role in Dordogne. Nine départements in environ-ments as diverse as the Mediterranean coast, the Loire Valley and the Gironde lost 500–1000 ha of vines annually, and in a further 22 départements between 250 and 500 ha were lost each year. Vineyards in 'marginal' locations, such as Lower Burgundy, Morbihan and the bas pays of Brive, all but disappeared (Moreau 1958). Ordinary wines produced in Upper Burgundy and Franche-Comté could not withstand cheap competition from the Midi, and mixed farmers throughout eastern France experienced special hardships because of the cereal crisis, which meant that they were simply unable to raise the necessary finance to reconstitute their vineyards (Claval 1978). In addition some vineyards, such as those of Isère which had been re-planted following phylloxera, later succumbed to the competition of cheap wine from new vineyards in the far south which came into full production soon after 1900 (Barral 1962, Ministère de l'Agriculture: Isère 1937). Only privileged vineyards, such as those in the Côtes de Champagne, parts of Burgundy and the Val de Loire, managed to survive in the northern half of France. A mere 2.9 per cent of the whole country remained under vines in 1908 and only Rhône (11.8 per cent), Gard (12.2 per cent), Gironde (15.1 per cent), Pyrénées-Orientales (15.2 per cent), Aude (18.5 per cent) and Hérault (30.3 per cent) devoted more than one-tenth of their surface to this land-use realm.

## Reconstitution

The botanist Planchon had determined the true cause of phylloxera as early as 1870, and experiments were initiated in the hope that a miracle cure might be found, which would prove as effective as sulphur had been against oïdium or copper sulphate against mildew. Application of insecticides allowed some phylloxerated vines to continue to produce, but this was no more than a costly half-measure which merely postponed the day of destruction. Trial and error in the far south showed that the spread of phylloxera might be prevented by planting vines in sandy soil which could be flooded in winter. Such an operation required land to be levelled or alternative means found to ensure that water could be evacuated as required. Action of this kind was only possible in a limited number of localities, such as the plain of Languedoc, where many *viticulteurs* organised themselves into syndicates to experiment with various ways of trying to combat the disease. The only effective measure was to rip out diseased French vines and replace them with resistant American strains, on to which French plants were grafted in order to preserve the traditional quality of the product. Removal of old vines, purchase of new plants, preparation of the soil, ploughing up and planting, together formed the complex and expensive process of reconstitution, which cost an average of 4000 fr/ha (Barral 1979). Unfortunately, the first returns on investment were not to be enjoyed until the new vines started to bear fruit in the fourth year after planting. Some financial assistance was made available by the state, départements and communes, but the greater share of the cost fell directly on landowners. Starting in 1887, newly planted vines were exempted from four years' land tax but, by this date, reconstitution was already underway on some holdings and had been judged impossible on others, since many small viticulteurs simply could not raise the necessary capital for such an adventure. By contrast, many large landowners, and especially those who enjoyed outside sources of finance, were able to respond much more readily (Galtier 1968). This structural contrast helps account for the decline of modest quality vineyards in many parts of France, and the emergence of large viticultural estates in Lower Languedoc which were organised along 'industrial' lines (Estier 1976). In addition, areas where viticulteurs established syndicates to cope with disaster, as in Languedoc and Beaujolais, managed to weather the storm more effectively than pays where vignerons had failed to organise themselves (Stevenson 1981, Garrier 1974).

Reconstitution was recorded specifically in the agricultural enquiry of 1892, when 442 295 ha of vineyard were declared to have been thus improved during the preceding decade, of which 84 per cent (372 610 ha) involved planting American vines (Stevenson 1978). Some 97 000 ha had been re-planted in Hérault and 73 700 ha in Aude, being followed by Gard (36 000 ha), Gironde (31 500 ha) and Pyrénées-Orientales (25 700 ha) (Hitier 1899). These five départements together accounted for three-fifths of all activity between 1882 and 1892. American vinestocks were used almost exclusively in Pyrénées-Orientales (98 per cent), Hérault (96 per cent), Aude (97 per cent) and Gard (95 per cent) and comprised more than 85 per cent in six other départements which fringed Languedoc to the north (Fig. 8.1c). 'Newly planted' vines had been recorded in 1882 when they amounted to 245 395 ha (11.4 per cent of all vineyards) and were most extensive in Aude (25 960 ha), Hérault (19 230 ha) and Gers, Lot-et-Garonne, Haute-Garonne and Tarn, each of which exceeded 10 000 ha. Vineyards in all of these districts had been devastated by phylloxera during the 1870s, and the six départements accounted for some two-fifths of all new planting. However, between 5000 ha and 10 000 ha were freshly planted in 10 other départements, especially in the western quarter of France, and smaller areas of new planting were recorded in every other viticultural département. Reconstitution undoubtedly accounted for a share of this trend, but completely new planting was also being undertaken by landowners who sought to capitalise on the misfortunes of the far south. Thus, in 29 départements, over 10 per cent of the vineyard surface was declared to be newly planted in 1882, with highest proportions in Gard (34 per cent) and Bouches-du-Rhône (35 per cent) (where phylloxera had struck as early as 1862) and in Vaucluse (23 per cent) and Hérault (28 per cent). Reconstitution predominated in these areas, whereas the principle of comparative advantage had promoted recent planting in Charente-Inférieure (38 per cent newly planted), four départements around Toulouse (each over 15 per cent newly planted), and many other parts of France.

By 1892, phylloxera had reached every viticultural département of any consequence and the distribution of newly planted vines was a clear response to that fact. However, the 301 225 ha of newly planted vineyards (17.2 per cent of the total) recorded in the statistics must have been a serious under-statement of reality, since none was mentioned in Hérault, even though much new planting had been cited in that département in 1882 and a great deal of activity occurred in neighbouring départements during the following decade (Fig. 8.1c). Most extensive new planting in the 1880s was recorded in

Aude (49 165 ha), and more than 5000 ha of post-phylloxera replanting occurred in other southern départements and in the more northerly locations of Rhône, Saône-et-Loire, Maine-et-Loire and Indre-et-Loire. By 1892, newly planted vines covered more than 35 per cent of all vineyards in five départements (especially Aude 42 per cent, Dordogne 43 per cent, Charente 45 per cent) and between 20 and 35 per cent in a further 17 départements (Fig. 8.1d). Advantage was taken to plant new vines in regularly spaced rows, so that mule-drawn ploughs might work the intervening spaces, while in Provence the old arrangement of vines interspersed with wide cultivated strips was replaced by higher densities of plants arranged in regularly spaced rows (Gay 1967, Masurel 1958). A veritable fever of re-adjustment not only followed phylloxera but even preceded it in areas which were spared until the two final decades of the century.

As a result, many vines had not started to produce at the time of the agricultural enquiries, amounting to 73 815 ha (30 per cent of all new vines) in 1882 and 56 450 ha (19 per cent) in 1892. Subsequent *enquêtes annuelles* did not refer to 'newly planted' vines as such, but information was gathered on surfaces that were 'not yet in production' and these data indicate that considerable replanting must have continued during the 1890s and after the turn of the century. In 1902, 172 840 ha (one-tenth of the national vineyard) had not yet started to produce, and in nine départements along the Loire and in middle France the proportion exceeded 20 per cent. Ten years later, the national total had fallen to 59 810 ha (3.8 per cent) with foci of late replanting standing out in Burgundy, the lower Loire, and parts of the Massif Central (where, of course, the viticultural surface was extremely slight). Eight départements contained more than 2000 ha of still unproductive vines, with peaks in Hérault (4930 ha), Gard (3560 ha) and Maine-et-Loire (3670 ha).

The spatial structure of French viticulture was undergoing dramatic changes as the new century dawned. New vines came into production, output soared to recall the yields of the 1870s, and the threat of massive surpluses reappeared (Fig. 8.3). Legislation, dating especially from 1897, reduced imports from Italy and Spain, but 3–5 000 000 hl of Algerian wine continued to enter France each year (Fig. 8.4). The price of vin ordinaire fell but serious frost caused a poor domestic harvest in 1903, which afforded a short-lived recovery for French viticulture; but the following year produced another massive harvest and prices tumbled once again. Coffee, beer and other beverages had gained popularity during the phylloxera crisis and this trend was to the detriment of the wine trade. In addition, part of the market for distilled alcohol had been captured by liquor derived from sugar beet, while the practice of adulterating wine raised additional threats to the future viability of French viticulture. Demonstrations and social unrest in 1907 made manifest the plight of many vinegrowers in the far south, whose once-rare commodity was now literally flooding on to the market in excessive supply (Napo 1971).

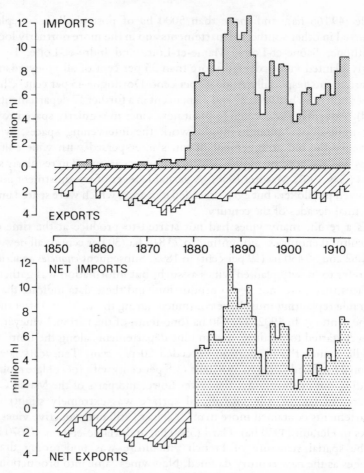

**Figure 8.4**   Wine, imports and exports, 1850–1912.

Only seven of the 71 départements which composed the viticultural realm contained more extensive vineyards in 1907 than at the time of the ancien cadastre (Fig. 8.1e). At the other extreme, 10 'marginal' départements had lost more than 90 per cent of their earlier total. In fact, all districts had undergone complex land-use changes during the intervening three-quarters of a century. One-third of the viticultural départements (26) conformed to the national trend for vine growing to gain ground until 1879 and then decline (++−) (Fig. 8.1f). Parts of middle France, the south-west and the Massif Central evolved in this way, as phylloxera and consequent reconstitution made their effects felt. In nine départements in Lower Burgundy, Vivarais and the Alps, areas under vines started to decline at mid-century (+−−), while vine growing retreated during all three phases (−−−)

in 17 départements in the Paris Basin, Franche-Comté, eastern Armorica, Bourbonnais and Limousin, where environmental conditions proved too difficult for viticulture to succeed at a time of increasing competition from the south (Ackerman 1977). Fourteen départements experienced a change in fortune between 1851 and 1879, as a phase of increase contrasted with earlier and later periods of decline $(-+-)$. These areas in the Loire Valley, Burgundy and Dauphiné had been spared from early attacks of phylloxera and had capitalised on the misfortunes of the far south to extend their vineyards. But such well-being proved short lived; phylloxera then spread northwards, many old vineyards in the Midi were reconstituted and others were planted anew during the final three decades prior to World War I (Viallon 1977). Bouches-du-Rhône, Gard, Hérault and Vaucluse had undergone expansion during the July Monarchy, devastation during the third quarter of the century, and impressive reconstitution during the Third Republic $(+-+)$; while the neighbouring département of Drôme shared in reconstitution after a protracted period of decline $(--+)$. To a greater degree than before, the well-being of the far south was to be bound up with the fluctuating fortune of the fruit of the vine.

# 9  *Devastation and conservation*

## Contrasts

Nineteenth-century observers expressed strongly contrasting views on the extent and condition of woodlands in France. A minority remarked how much timber survived in some parts of the country, with Colman (1848) being 'constantly impressed by the immense tracts of land in forest' and describing 'arks of charcoal and wood' that were floated down the Seine and were then gathered into great piles at various points alongside the river in Paris (p. 23). Undoubtedly some pays still supported great expanses of trees, like the 'immense pine forests' of the Landes where 'the presence of man is virtually unknown', but more devastating descriptions were far more usual (Mugriet 1821, p. 8). Woodlands that had been formerly owned by the Crown, the nobility, the Church and even by peasant communities had come under widespread attack after 1789, reducing both the wooded surface and the quality of the trees that survived. Large numbers of hungry livestock overgrazed extensive areas and timber remained in great demand for construction, fuel and manufacture of implements. For example, the forests of the upper Saône valley were being felled to supply shipbuilders as far afield as Toulon on the Mediterranean coast, while woodlands in Burgundy served Paris and shipyards along the lower Loire, as well as local centres of demand on the Rhône–Saône axis (Richard 1978). Regulations that were intended to control the exploitation of forest resources were abused almost universally, with wise rotations for thinning and felling being observed in precious few areas. The temptation to clear trees and grow crops was even greater when grain was in short supply and cereal prices were high. In this way, Bellecombe forest in Haute-Loire was felled in the immediate wake of the high cereal prices of 1817–18 (De Sainte Colombe 1828). Examples of this kind of response were both legion and widespread during the first half of the 19th century (Cotheret 1836, Maury 1976).

Indeed, the woodlands of most mountainous districts were in a sad state at the dawn of that century, as in Hautes-Alpes where Prefect Bonnaire reported in 1798 that 'beautiful forests' in Briançonnais were being cleared to create 'the most terrible nudity . . . of bare and sterile rocks' (cited in Crubellier 1948, p. 279). In addition, the widespread practice of systematically burning shrubs and small trees not only had the desired effect of increasing pasture space but also prevented the regeneration of woody vegetation, since seeds and roots were destroyed as well as surface vegetation. In this way, sections of the Causses and other southern pays had been converted into 'landscapes of desolation' (Marres 1942, p. 180). Conditions were possibly worst in Provence where woodlands were 'becoming rarer every day' and 'were no longer worthy of the title "forests"' (Villeneuve & Robert 1839, p. 62). Woodcutters were poorly trained and failed to protect trees from casual devastation by local inhabitants, who not only cleared vegetation to obtain kindling but also removed rotting leaves and top soil to use as compost on their fields. Small flocks of sheep finished off remaining traces of vegetation, with areas alongside transhumants' drove-roads often being totally devastated. Firewood was virtually unobtainable in much of Provence in the early 19th century and the peasants of some mountain districts took to moving in with their livestock or burning cow dung as they attempted to keep warm in winter. Many commoners actually started forest fires as a way of increasing grazing land for their sheep and were, of course, 'totally opposed to afforestation' (Puvis 1845–6, p. 202). Not surprisingly, most communal woodlands in Provence were denuded of trees and shrubs, which made them easy prey to soil erosion when torrential mediterranean rainfall occurred. Only one-tenth of the so-called woodland in Var gave regular yields of timber during the July Monarchy; the remainder was covered with scrub. Baudrillart's (1831) memoir on deforestation reported similar disasters throughout the Jura, Pyrenees and Alps; and Purvis (1839–40) estimated that the tree cover in the Vosges had declined by half between 1789 and 1839.

Intelligent observers such as these were well aware of the long-term ecological implications of this devastation, as they recorded hillside erosion, newly exposed areas of naked rock, and once-fertile valley floors that had become strewn with stones and gravel. It was even claimed that some settlements in the mountains of Cantal had been abandoned recently because of accelerated soil erosion and a total lack of timber, while further north the plateaux of Champagne formed a 'land of desolation on which trees and shrubs were rarely to be seen' (De Chabriol 1822, Bosc 1826, p. 60). Cereal yields declined devastatingly after only a few years of cultivation in areas which had been freshly cleared of timber in Côte-d'Or, and identical problems were encountered on poor soils throughout France (Ministère de l'Agriculture: Côte-d'Or 1937). By the 1820s and 1830s, administrators were advocating that softwoods should be planted not only in the uplands but also in low-lying mauvais pays, such as the brandes of Indre where infertile soils seemed 'to offer no advantage apart from growing trees' (Moll 1838–9, p. 9).

Attempts had been made to encourage afforestation in the late 18th century but results were slight. For example, during the 1790s Brémontier had been awarded a medal by the Société d'Agriculture de la Seine, in recognition of his effort to anchor shifting dunes in Gironde through planting reeds and trees (Arqué 1935). A Commission des Dunes was established under his direction to pursue this objective more vigorously, and duly operated until 1817 when the work was conferred on the administration of the Ponts-et-Chaussées. In addition, Prefect Haussez was attempting to encourage communities in Gironde to divide communal wasteland as a preliminary to afforestation, but his efforts were to no great avail (Larroquette 1924). Further north, a number of estate owners in the Sologne and on the chalky steppes of Champagne had started to plant softwoods (especially Scots pine) during the 1790s and again after the fall of the First Empire (Gillardot 1972, Hau 1976). For example, in 1815 a certain Monsieur Saint-Denis had planted timber on 5 ha of wasteland in Marne; his sons continued the good work and by the end of the July Monarchy had planted up 4000 ha of their own land and more than 12 000 ha that belonged to others (Huffell 1904). Their efforts were emulated by surrounding landowners, and trial and error showed that Austrian pines were not only more suited to poor calcareous soils but were also more resistant to disease. Extremes of drought and cold caused great harm to experiments such as these, but a small start had been made on planting up the mauvais pays of lowland France and this urgent task was to continue dramatically after mid-century.

Many highlanders expressed strong opposition to the idea of afforestation and even to schemes for more efficient management of existing woodlands, since they feared that their resources of rough grazing land, poor though they were, might disappear beneath a mantle of trees. The forest ordinance of 1669 remained the basis for woodland administration in the early 19th century, even though legislation of 1791 had intimated that new forest laws would be forthcoming. Preparation for new legislation was started in earnest in 1822 and, two years later, a national forestry school was created at Nancy (Reed 1954). In 1827 the code forestier was passed which stipulated that public forests, whether owned by the state, communes or other public bodies, would henceforth be managed according to a special regime (Dion 1970). State forestry officials were to mark the outer limits of the forests more clearly; industrial enterprises and sawmills were forbidden in their immediate surroundings; regulations for exploiting timber and for grazing livestock were to be enforced strictly by foresters (who were nominated by state officials and not by the inhabitants of local communes); and offences committed in public woodlands were liable to incur fines or imprisonment. Attempts by officers of the Eaux-et-Forêts to implement the new code provoked fierce hostility and contributed to the guerre des Demoiselles of 1829 in Ariège, which was followed by outbreaks of popular opposition elsewhere in the Pyrenees, Alps and Jura. In brief, well meaning forestry regulations continued to be flouted in many parts of France throughout the 19th century (Guillard 1980). Livestock were still driven into communal woodlands as they always had been, regardless of likely ecological consequences or the stipulations of the new regime (Henry 1943). Many foresters connived with commoners, or at least managed to turn a blind eye. Veritable frondes ensued when the law was strictly enforced and the militia sometimes had to be summoned to quell such disturbances.

A series of official reports into the causes of serious flooding during 1840

depicted the lamentable state of woodland in most mountain areas and explained the relationship between déboisement, accelerated runoff, intensified
erosion and flooding (Brown 1876). Surrell recorded such disasters in the
southern Alps, and his message was reinforced by Adolphe Blanqui's (1846)
wide ranging report to the Institut de France, entitled *Du déboisement des
montagnes*. He argued that many districts in the high Alps already formed 'a
sort of Arabia Petraea' and he feared that unless emergency legislation were
passed to start afforestation 'in 50 years there will be another such desert
between France and Piedmont as there is between Syria and Egypt' (p. 72).
Planting of trees was duly initiated in trial areas and the afforestation of
mountain areas was encouraged by new legislation in 1860 and later years.
Each successive disaster, such as the floods of 1875 in the Massif Central or of
1890 in Ardèche, demonstrated the urgency of such an enterprise (Huffel
1904).

## A narrow view

Cadastral data are less helpful for examining woodland than they are for
other land-use realms, since state-owned property was exempt from
cadastral levy. Hence, alternative statistical sources need to be employed
which take account of national forests in order to determine the total woodland surface. The cadastral record shows the national woodland cover
increasing from 7 673 555 ha in the ancien cadastre to 8 216 135 ha in the
revision of 1907; and when state woodland is included the total is raised from
8 572 850 ha early in the July Monarchy to 9 257 580 ha on the eve of World
War I (Table 4.1). Both the cadastral and composite data sets reveal similar
trends during the intervening decades, with a slight reduction in woodland
occurring between the 1830s and mid-century, which was followed by
increases in subsequent phases. Cadastral information indicates that the
national total of private and communal woodland increased by 542 580 ha
(up 7.1 per cent) between the ancien cadastre and 1907. In fact, only 38
départements registered a net increase in tree cover and were located along a
diagonal from Landes, through Limousin, middle France and Champagne,
to the Ardennes, with offshoots in Auvergne and Franche-Comté and an
important outlier in Provence (Fig. 9.1a). Gironde (up 97 per cent), Landes
(up 83 per cent), Marne (up 72 per cent) and Haute-Vienne (up 57 per cent)
recorded the largest proportionate increases, whereas the greatest volumes of
net afforestation occurred in Landes (up 220 850 ha), Gironde (up 183 900
ha), Marne (up 67 360 ha) and Dordogne (up 64 420 ha). The remaining
44 départements contained less cadastral woodland in 1908 than three-
quarters of a century earlier and were arranged in two distorted crescents;
one extending from Flanders through Picardy, Normandy, Armorica and
the Charentes, and the other running from Nivernais, along the Saône and

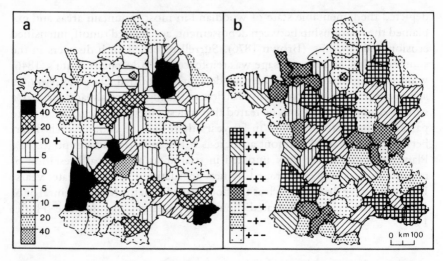

**Figure 9.1** All woodland: (a) net change in surface, ancien cadastre–1907 (per cent); (b) typology of change.

Rhône, through the southern Massif Central to the Pyrenees. Largest proportionate losses occurred in Manche (down 25 per cent), Pas-de-Calais (down 30 per cent), Loire-Inférieure (down 34 per cent), Corrèze (down 57 per cent) and Seine (down 87 per cent); with Corrèze losing 64 610 ha of woodland cover, Basses-Pyrénées 29 810 ha and net losses of 10–25 000 ha affecting 15 more départements in Picardy, Normandy and southern France.

Cadastral woodland declined by just 20 910 ha over France as a whole during the July Monarchy but increases were recorded in 27 départements at that time, with the figure rising to 37 between 1851 and 1879 and to 52 after 1879. A dozen départements, mainly in Picardy and northeastern France, lost over 250 ha of woodland each year during the first phase, with the largest losses being in Aisne (down 590 ha p.a.), and Vosges (down 695 ha p.a.) (Fig. 9.2a). Extensive areas of land were being put under the plough in virtually all of these départements during the July Monarchy. Ongoing demands for timber and the profitability of sugar beet and other new crops combined to generate this trend in many northern districts (Fossier 1974). Only Loir-et-Cher (up 575 ha p.a.), Marne (up 763 ha p.a.) and Hautes-Alpes (up 4240 ha p.a.) increased their cadastral woodland by more than 500 ha p.a. at this time, with wasteland forming the main feature to decline in the first and third départements, and ploughland undergoing substantial retreat in Marne (Fig. 9.2b). Between 1851 and 1879 cadastral woodland advanced by 273 360 ha, with truly massive net increases occurring on former areas of wasteland in Gironde (up 4765 ha p.a.) and Landes (up 8345 ha p.a.) (Fig. 9.2c). Eight other départements in eastern and middle France registered increases of more than 250 ha p.a., with important advances in Marne (up 940 ha p.a.) and

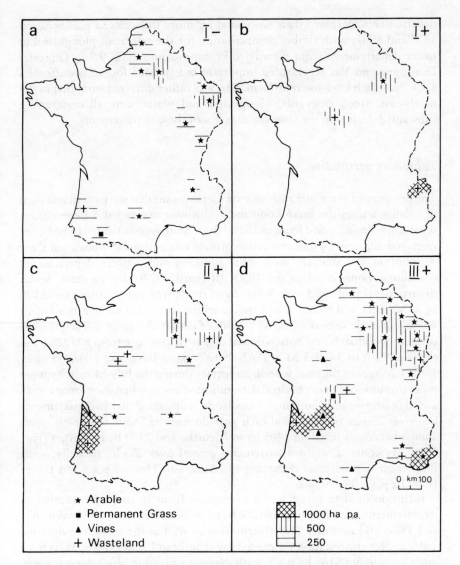

**Figure 9.2**  Cadastral woodland, annual net change by cadastral phase.

Aube (up 715 ha p.a.). Arable land retreated markedly in both of these départements but wasteland was the main loser in other regions. From 1879 to 1907 cadastral woodland rose by a net total of 290 130 ha, namely at the same rate as in the previous phase. In fact, the local rate of increase had slackened even in the leading départements (e.g. Gironde up 1780 ha p.a., Dordogne up 1915 ha p.a.), however the process of afforestation had become much more widespread than in earlier phases. No fewer than 25 départements increased their woodland by more than 250 ha p.a. between 1879 and 1907, with timber compensating for the retreat of ploughland in many départements from Picardy to Franche-Comté (Fig. 9.2d). Gironde, Dordogne and Var represented important but isolated foci of net afforestation, in which land-use transformation was rather different from that in the north-east, since vineyards, ploughland and waste were all undergoing substantial decline in the land-use mosaics of these départements.

## A broader perspective

The presence of extensive areas of state forest meant that the pattern and pace of change within the total woodland realm was somewhat different from what has been identified from cadastral data. State forests tended to be better managed and more productive than private or communal woods but their distribution was uneven, with some southern and western départements containing none at all in the 1830s (Richard 1927). By contrast, seven départements possessed over 30 000 ha of state forest apiece, rising to 72 105 ha in Meurthe and 86 225 ha in Vosges, which accounted for one-third of all woodland in these two areas. As a result of net afforestation and acquisition of other woodlands, the national total of state forest rose from 899 325 ha in the late 1830s to 947 985 ha in 1862, 952 370 ha in 1882, and 1 041 445 ha in 1907. Changes in the total woodland surface during the July Monarchy were more pronounced than cadastral data indicated since, when state forests were added to private and communal woodlands, it appeared that 18 départements lost over 250 ha of woodland each year during the 1830s and 1840s, with annual net losses reaching 1405 ha in Meurthe and 2370 ha in Vosges (Fig. 9.3a). By contrast, eight départements gained over 250 ha annually, with the largest net increases occurring in Marne (up 1180 ha p.a.) and Loiret (1700 ha p.a.) (Fig. 9.3b).

Inclusion of state forest serves to increase from 10 to 11 the number of départements gaining more than 250 ha of woodland each year between 1851 and 1879, and modifies their distribution as well as the absolute values involved. Net afforestation proceeded vigorously in Landes (up 9035 ha p.a.) and Gironde (up 5165 ha p.a.), with extensive planting also taking place in Champagne, the Pyrenees, the Alps, Vosges and the Massif Central (Fig. 9.3c). In the quarter century preceding 1859, 320 620 ha had been felled and only 279 520 ha planted, producing a net loss of 41 100 ha; but when the span of years is advanced to 1866 the respective values are transformed to 421 885 ha and 520 440 ha, to give a net increase of 98 355 ha. Thus, it would seem that the national trend changed abruptly in the middle years of the Second Empire, to enable the woodland total to advance by 139 655 ha between 1859 and 1866, as a result of planting 240 920 ha and felling 101 265 ha. Unlike the impression conveyed by examining net changes, it is clear that woodland

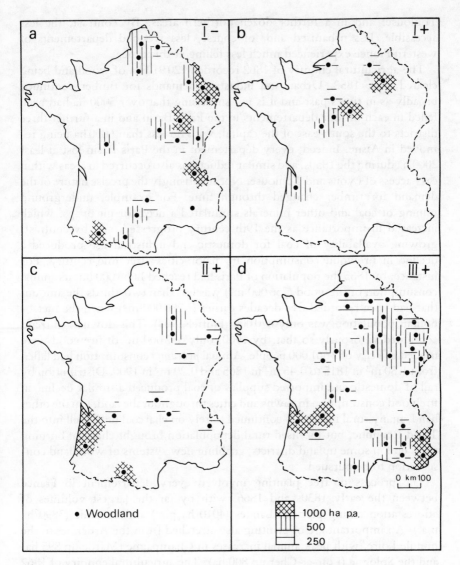

**Figure 9.3** All woodland, annual net change by cadastral phase.

was actually being felled in every single département between 1830 and 1866, with more than 400 ha being lost each year in a block of districts located between Paris and the major demand centres of the north (peak loss of 620 ha p.a. in Aisne), and in the outlier of Saône-et-Loire which contained industrial towns of its own and provided supplies to the city of Lyons (De Chambray 1834). Between 200 and 400 ha were felled each year in a scatter of départements in Burgundy, along the lower Rhône, around Bordeaux and

Toulouse, and in a further dozen northern areas. By contrast, the less accessible, less urbanised and, of course, less wooded départements of western France experienced much less felling at this time.

The agricultural enquiry of 1862 recorded 120 910 ha of woodland being cleared since 1852. Urban and industrial demands for timber continued broadly as in times past and it is not surprising that over 5000 ha had been felled in each of four départements in the Paris Basin and in a further three districts to the south-east of the capital, with no less than 6000 ha being removed in Aisne. Indeed, every département in the Paris Basin lost at least 2000 ha during the 1850s, and similar reductions also occurred in areas within easy access of Lyons and Toulouse. Not surprisingly the precise nature of the demand for timber changed through time. For example, underground mining of coal and other minerals stimulated a need for pit props, which increased in importance as the 19th century progressed, and by contrast, growing availability of coal for domestic and industrial use produced a decline in firewood consumption in major cities after mid-century. For instance, by 1865 the population of Paris had reached 1 668 000 but its annual consumption of firewood (756 000 m³) was less than two-thirds the amount that had been recorded five decades earlier (1 200 000 m³) when the metropolitan population was only 670 000 (Huffel 1904). This downward trend continued vigorously so that, by 1900, only 552 000 m³ of firewood were needed in a city of 2 661 000 people. Annual *per capita* consumption had fallen from 1.80 m³ in 1815 to 0.45 m³ in 1865 and 0.20 m³ in 1900. Distribution by rail of domestic and imported supplies of coal produced a similar decline in firewood consumption in towns and cities throughout the land. On the other hand, many rural industries continued to rely on charcoal until well into the Third Republic, but localised rural depopulation brought charcoal burning to an end in some upland districts, enabling new systems of woodland conservation to be pursued.

The process of tree planting involved every département in France between the early 1830s and 1866, with by far the largest volumes of afforestation taking place in Landes (4010 ha p.a.) and Gironde (3890 ha p.a.). An important arc of planting also stretched from the Ardennes to the middle Loire, with peak annual increases in Champagne (Marne up 895 ha) and the Sologne (Loir-et-Cher up 800 ha). The agricultural enquiry of 1862 cited 129 060 ha of tree planting since 1852, with only Cher and Deux-Sèvres recording no activity during that decade. By far the most widespread afforestation occurred in Landes (61 741 ha) and Gironde (24 060 ha) and over 2000 ha being planted up in a scatter of départements in and around the Massif Central. At the other extreme, very little was accomplished in the Alps, Franche-Comté or Normandy at this time.

Between the cadastral survey and 1866 no fewer than 52 départements experienced a net decline in their woodland surface, with a total of 78 500 ha being lost in five départements in Flanders and Picardy (peak loss of 19 235 ha in Aisne), and Meurthe (down 13 755 ha) and Saône-et-Loire (down 11 125 ha) also registering important losses. With the exception of Champagne, all parts of the Paris Basin contained less woodland in 1866 than in 1830, and the same fate characterised départements in Burgundy, along the Rhône, around Toulouse, and at the mouth of the Loire (Brosselin 1977, Sornay 1934). A similar pattern was recorded in the statistics for 1852–62, with felling exceeding planting in no less than 61 départements and net losses of more than 4000 ha occurring in seven départements to the north and north-east of Paris. In addition to the commercial demand for timber, forest fires caused serious losses in the arid Midi and especially in the Maures and Esterel massifs, where carelessness by hunters, charcoal burners and shepherds easily promoted disaster, as accumulations of dry leaves, heather and brushwood caught light and in turn caused more substantial trees to ignite (Ministère des Finances 1869). Tighter enforcement of forestry codes was being advocated strongly toward the end of the Second Empire, as was the cutting of new roads into isolated areas of Provence where forest fires had been able to rage completely out of control. The Landes of Aquitaine formed by far the most extensive focus of net afforestation during the July Monarchy and the early part of the Second Empire, with Gironde and Landes together increasing their woodland surface by 247 000 ha in the course of 35 years. The départements of Aube and Marne in Champagne augmented their woodlands by 34 800 ha; while in middle France the combined increase for Loir-et-Cher, Indre-et-Loire and Indre reached 33 700 ha. Returns for 1852–62 confirm the supremacy of Landes and Gironde, where woodland increased by 84 500 ha in only 10 years, but indicate net reductions in Champagne and middle France.

## Transformations

Afforestation and drainage of moors and marshes in Aquitaine had been set on a new footing in 1842, when a young Ponts-et-Chaussées engineer named Chambrelent was appointed to Gironde. He promptly acquired 500 ha of marshy land for improvement and was so successful that his experiments were emulated by other landowners, enabling 20 000 ha of wasteland to be put under trees by 1855 (Mangon 1860). Legislation two years later encouraged local authorities to drain land, construct agricultural roads and plant commons with trees. However, traditional rights of intercommoning hindered improvement across wide stretches of land, and were at the root of bitter conflicts of opinion between neighbouring communities on the subject of land-use change (Papy 1978). For example, communal shepherds continued to insist that it was necessary to burn rough vegetation to produce adequate pasture for their sheep, even though this practice was totally at odds with afforestation.

The American Civil War cut off imported supplies of resin into France and, in 1861, enabled landowners in Aquitaine to appreciate the economic value of the new forests and to begin to make effective use of these resources. Spurred on by such events, afforestation continued apace during the Second Empire and the Third Republic, enabling the 'kind of interior Sahara' that Chambrelent (1887, p. 21) had described to be converted into a vast but vulnerable sea of pines. Unfortunately, forest fires brought localised destruction at this early time as they were to do in later years. Ancient woodlands and well policed areas of state forest experienced relatively little damage but stretches of new pines in either private or communal ownership proved particularly liable to fire damage. Sparks from railway engines, stray matches thrown down by hunters, and the malevolence of shepherds all came in for blame. In order to minimise damage, officials advocated greater surveillance and cutting of 20 m firebreaks to either side of railway tracks (Ministère des Finances 1873).

The Champagne pouilleuse occupied 700 000 ha in a vast quadrilateral of chalkland stretching 160 km between Rheims and Sens, and 45 km between Epernay and Vitry-le-François. At the beginning of the 19th century, some 500 000 ha of *savarts* remained as rough grazing, with occasional patches of temporary cultivation being found in less impoverished areas (Dion 1961). The greater part of this land was duly put under the plough during the course of the 19th century, but afforestation offered a more appropriate form of land use on very poor soils, with Austrian black pine proving more successful than other varieties. The communal status of the savarts retarded changes in some localities but planting gradually occurred, so that as increasing supplies of firewood became available, the ancient practice of burning dried dung as fuel started to disappear (Le Play 1879).

Afforestation also formed an integral part of the improvement schemes that were implemented in the mauvais pays of middle France. For example, the Comité Central Agricole de la Sologne, under the chairmanship of the Directeur Général des Forêts and with the collaboration of *savants* such as Elie de Beaumont and Michel Chevalier, produced a programme of widespread afforestation for the second half of the 19th century. Maritime pines were planted over extensive areas of the Sologne, but this good work was brought to a brutal halt in December 1879 when many young trees perished in an intensely cold spell. With help from state foresters and aided by official grants, local landowners started to replant immediately but this time they made use of Scots pine, which had already proved successful in some national forests in middle France.

Between 1879 and 1907 the total amount of land under trees increased by 379 205 ha, a figure almost identical to that cited for 1851–79. Woodland advanced in 52 départements and by more than 250 ha p.a. in no less than 31 of these (Fig. 9.3d), by comparison with 25 when state forests are excluded from the calculation (Fig. 9.2d). The whole of northeastern France was increasing its woodland surface very substantially and a change in emphasis was occurring in the south-west, where extensive planting of softwoods in Dordogne (up 1985 ha p.a.) followed the onslaught of phylloxera, and enabled that département to overtake both Gironde (up 1525 ha p.a.) and Landes (down 350 ha p.a.). An important new focus of net afforestation appeared in Var (up 4440 ha p.a.) and Basses-Alpes (up 1250 ha p.a.). Only six départements were losing more than 250 ha of woodland each year during this phase, with annual losses being particularly extensive in Basses-Pyrénées (down 1085 ha p.a.), Ariège (down 2235 ha p.a.) and Corrèze (down 2445 ha p.a.).

By contrast with the relative abundance of information relating to woodland in the Second Empire, much less is known about precise processes of change during the Third Republic. The only statements were made for the years between 1882 and 1892, when 47 185 ha were declared to have been felled and 114 115 ha planted. The amount of woodland cleared at this time was less than half that recorded for 1852–62 but total planting was only fractionally less. Every département in the land experienced some felling during the 1880s but the pattern was very different from that at mid-century. During the intervening decades, coal had replaced timber as a source of fuel in the cities of northern France and that fact, together with the diminishing tree cover of Picardy, meant that the northern focus of felling had disappeared completely. Between 1882 and 1892, peak rates of clearance were recorded in Orne (down 3340 ha), Gironde (down 3050 ha) and six contiguous départements between Loiret and Loire, in which more than 1000 ha were felled apiece. Major concentrations of planting activity continued to be found in Champagne (Marne up 5790 ha), middle France (Loir-et-Cher up 8350 ha, Loiret up 6560 ha) and Aquitaine (Gironde up 6975 ha), with important new activity appearing in the south-east, north-east and Limousin. Between 1882 and 1892, felling outweighed planting in only 25 départements.

By the eve of World War I, three-quarters of a century of clearance and afforestation had served to raise France's total woodland realm by 684 700 ha, which was roughly equal to the total surface of Loire-Inférieure or Nièvre. Net increases occurred in 40 départements but in only nine did they exceed 25 000 ha apiece. The mauvais pays and the mountains formed the main foci of net afforestation, with impressive achievements in Aquitaine (Landes up 239 925 ha, Gironde up 198 230 ha, Dordogne up 63 985 ha), Champagne (Marne up 64 775 ha, Aube up 41 295 ha), middle France (Loir-et-Cher up 35 805 ha, Loiret up 34 185 ha) and Provence (Var up 126 290 ha, Basses-Alpes up 41 375 ha) (Douguedroit 1976). On the other hand, half the départements of France supported less woodland in 1907 than early in the July Monarchy, with notable clusters of decline in the Pyrenees and along their northern margins, in southwestern parts of the Massif Central, and in central-eastern and northeastern France (Fig. 9.1a). In absolute terms, such reductions normally involved quite small areas but they surpassed 25 000 ha in Basses-Pyrénées (down 29 880 ha), Meurthe (down 43 575 ha) and Corrèze (down 65 960 ha).

The complex changes that produced these results may be summarised by the familiar typology of land-use evolution (Fig. 9.1b). Fifteen départements, located in Champagne, middle France, Aquitaine and Provence, increased their woodland during each cadastral phase (+++). These foci were expanded by a further 13 départements in which woodland evolved in concert with the national trend, whereby the tree cover declined during the July Monarchy but increased during both subsequent phases (−++). In eight départements of northern France, the trend of growth was interrupted during the second phase (+−+); while five dispersed départements were characterised by two phases of expansion, giving way to decline in the final years of the 19th century (++−). The remaining 41 départements were dominated by negative trends, with Lower Normandy, Gascony, the hinterland of Lyons and the heart of the Ile-de-France display-ing losses throughout the period (−−−). Three scattered areas underwent decline after mid-century (+−−), while 10 départements in the Alps, Pyrenees and Massif Central experienced a single phase of expansion during the third quarter of the 19th century (−+−). The remaining 16 départements, located in the Paris Basin, the north-east, Burgundy and the southwestern corner of the Massif Central, managed to increase their wood-land cover after 1879 following half a century of decline (−−+). Each of these developments produced important changes in rural landscapes throughout France, but nowhere more profoundly than in the Landes of Aquitaine which Daniel Zolla (1920) could describe as being 'metamor-phosed by trees' (p. 20). Health conditions had improved greatly in each of the interior colonies and the threat of further environmental degradation had been brought into some degree of check by judicious afforestation in many mountain localities. Nonetheless, problems of excessive runoff had not been completely mastered, and devastating floods were still causing fatalities in the Pyrenees and other highland regions at the close of the 19th century; while forest fires continued to represent serious threats to life and landscape alike in the Landes and Provence (Gaussen 1932).

# 10    *Sum of the parts*

*Only connect*

Despite the strength of peasant mentalités and a continuity of many aspects of rural life from earlier times, the peaceful century between the Napoleonic Wars and the outbreak of World War I encompassed an intricate transition from the ancien régime économique to the modern state of France. The discipline of distance was partially rewritten following the construction of railways and routeways and, as a result, many farmers took advantage of directing a share of their production to satisfy market demands rather than simply trying to subsist. New systems of connectivity were paralleled by important changes in population numbers with the national total increasing by one-third and the urbanised proportion exceeding two-fifths of that figure shortly before World War I. Thus in 1911, 44 per cent of France's 38 455 000 inhabitants lived in 'urban' communes (with settlement clusters of more than 2000 people) and 38 per cent resided in towns with over 5000 inhabitants. Nonetheless, France lagged behind her European neighbours with respect to total growth and urbanisation. At the focus of the transport system and of national life, Seine département housed 4 154 000 people and Marseilles, Lyons and Lille (together with their suburbs) each surpassed the 500 000 mark (Gras 1979). These four centres contained 15 per cent of the French nation and the next six major towns housed a further million people, thereby accounting for one-fifth of the country's food consumers. The rise of these and other towns was achieved at the expense of the countryside, as townward migration and subsequent demographic devitalisation in rural areas took their toll. As a result, only a score of départements were still increasing their population in the early 20th century and each of these contained at least one dynamic urban centre.

On the eve of World War I, Paris drew foodstuffs to its central markets from a truly national hinterland. Fruit and vegetables were rushed by train to the capital, with producers in each département taking advantage of their distinctive environmental conditions to specialise in commodities that would sell in season (Demangeon 1928, May 1930). In the far south, Hérault and Var together accounted for 23 per cent of the total weight of fruit handled in the Halles Centrales in 1913, and Lot-et-Garonne and Tarn-et-Garonne provided a further 19 per cent (Fig. 10.1a). No less than 28 per cent of the volume of vegetables marketed in Paris came from Var, with neighbouring Bouches-du-Rhône sending a further 12 per cent (Fig. 10.1b). Manche supplied 13 per cent and led the field in northern France, while the market gardens and vegetable plots of Seine-et-Oise contributed 7.3 per cent. Two-

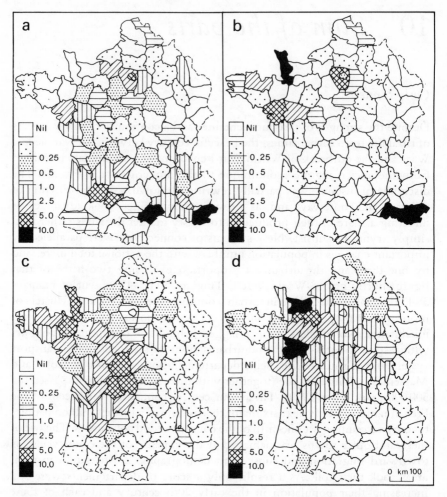

**Figure 10.1**   Origin of (a) fruit, (b) vegetables and (c) meat sent to Halles Centrales, Paris, 1912 (per cent). (d) Origin of cows sent to La Villette, 1912 (per cent).

thirds of the butter handled in Parisian markets came from the Charentes, with Brittany supplying a further 16 per cent. Other commodities were dispatched shorter distances: 42 per cent of the eggs marketed in Paris originated in Beauce and a further 40 per cent in Normandy; 44 per cent of the mushrooms were produced in Seine-et-Oise and 38 per cent in Seine itself; the valleys of Seine-et-Oise (52 per cent) and Oise (38 per cent) supplied almost all the watercress that Parisians consumed; while the capital's 'milkshed' extended into many sections of the Paris Basin (Billé 1930, Dubuc 1938).

Meat arrived from considerably more distant locations, with all départements save Pyrénées-Orientales and Var dispatching a share of the 44 115 000 kg of carcases sold in Parisian meat markets (Fig. 10.1c). Western and central France formed major points of origin, with Manche and Ille-et-Vilaine together dispatching 7 320 000 kg and Creuse, Indre and Haute-Vienne sending a combined total of 9 000 000 kg (Dulac 1900). In addition, 2 241 000 head of livestock were sent for slaughter at La Villette, comprising 1 360 000 sheep, 371 000 pigs and 510 000 cattle of various types. A predominantly northwestern origin characterised all supplies of beefstock but there were significant local differences according to types of cattle. Calvados (16 110) and Maine-et-Loire (12 310) together provided 27 per cent of the 105 000 cows sent for slaughter in Paris (Fig. 10.1d). Some 213 000 bullocks were sent to La Villette and the same two départements together accounted for no less than 59 130 but the zone of dispatch was rather more diffuse, extending from Calvados southwards through Maine, Anjou, Vendée and Poitou, with an outlier in Nivernais (Fig. 10.2a). The 167 000 calves sent to Paris also originated mainly in départements to the west of the capital, with Sarthe (36 500) and Eure-et-Loir (19 000) accounting for the largest share (Fig. 10.2b). By far the greatest proportion of the 370 000 pigs sent to Paris came similarly from the north-west, with Sarthe (15.6 per cent) and Loire-Inférieure (13.2 per cent) forming the most important foci within a wider zone which embraced Upper Normandy, Anjou and Vendée (Fig. 10.2c). Some 1 360 000 sheep were sent to La Villette, and all but eight départements sent some animals to Paris; however, the main foci of dispatch were quite different from those identified so far, comprising the Ile-de-France, Cantal, Tarn and surrounding départements in the southern Massif Central (Fig. 10.2d).

These commodity flows to the markets and slaughterhouses of Paris provide good illustrations of how the capital was connected to distant regions for food supplies, and of where distinctive specialisms had developed once the necessity for farmers in each locality to attempt to produce enough grain to survive had been removed.[1] The freedom whereby many farmers directed their activities to satisfying market needs was a clear response to government policy for railway construction and this development was further bolstered by commercial arrangements which enabled cheap grain to enter France during the 1860s and 1870s, thereby depressing cereal prices and offering new impetus for livestock husbandry. This trend was accentuated even further by an accelerating demand for meat commensurate with improvements in living standards as the 19th century elapsed. In short, once the railway enabled the discipline of distance to be mastered, farmers could start to exploit regional differences in resource endowment, farm size and landownership in whatever ways they perceived to be both appropriate and viable.

Of course, the transformation was far from automatic, simple or uniform, since traditions and profound beliefs long held by countryfolk had to be

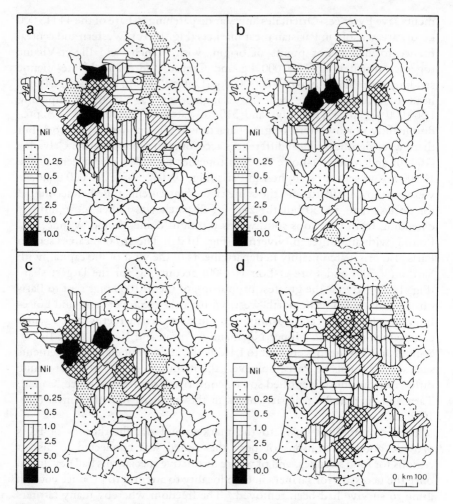

**Figure 10.2** Origin of (a) bullocks, (b) calves, (c) pigs and (d) sheep sent to La Villette, 1912 (per cent).

refashioned in the light of knowledge of the world beyond the village and by the diffusion of new ideas. This was gradually made possible by the combined impact of primary education, increased literacy, circulation of newspapers, development of more personal travel, and the process of mass conscription which took young men away from their farms and into urban barracks at a particularly crucial phase in their lives (Weber 1977). Despite the existence of powerful forces for maintaining continuity with the past, substantial agricultural changes did result. For example, the number of livestock units raised in France increased by a quarter between the 1830s and 1912 and the pattern of animal husbandry was modified significantly.

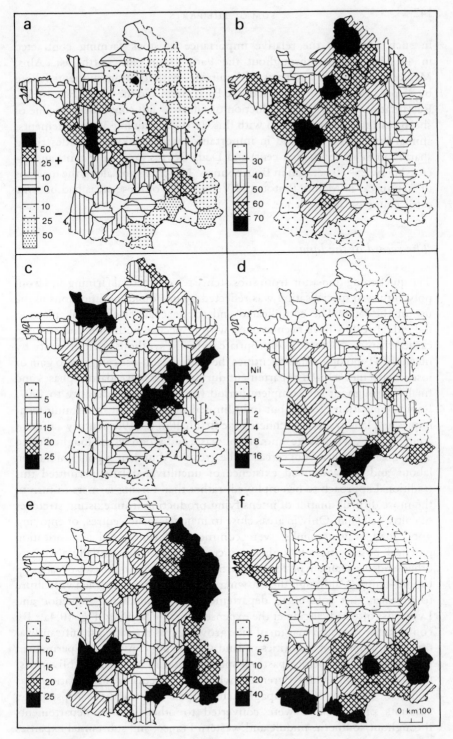

**Figure 10.3** (a) Change in location quotients for livestock units, 1837–1912 (per cent). (b) Arable, (c) permanent grass, (d) vines, (e) all woodland and (f) wasteland as percentage of total land use, 1907.

In such a context the relative importance of stock farming contracted in 44 départements throughout the Paris Basin, the north-east, Alps, Mediterranean south and parts of Aquitaine, with location quotients falling by 25 per cent or more in much of the Paris Basin and the south (Fig. 10.3a). Reclamation of wasteland, reduction of fallows, and consequent decline of sheep rearing had much to do with this. In the remaining 38 départements, animal husbandry increased in importance, with stocking densities rising sharply in Seine (up 62 per cent) and Deux-Sèvres (up 51 per cent) and increasing by 25–50 per cent in Bourbonnais, Nivernais and along the margins of Armorica, in Maine, Poitou and Vendée. All of these areas had become cattle country par excellence.

## The record of the land

The quickening transition from subsistence to commercial farming and from polyculture to specialisation was reflected by profound modifications in the use of the land. The complicated and subtle changes that took place in individual realms have been evoked with respect to both time and space in earlier chapters. Now it is appropriate to try to synthesise that diversity by means of a typology which identifies the leading land-use category to gain or lose importance in each département during successive cadastral phases and finally throughout the complete period from the ancien cadastre to 1907. Rapid population growth but only limited improvement in farming techniques and land transport rendered the first half of the 19th century a time of 'land hunger', when large amounts of waste and wood were ploughed up to allow greater quantities of cereals to be grown. Abundant supplies of human labour and the continued existence of uncultivated land permitted this traditional approach to be employed, rather than forcing farmers to turn to the more difficult matter of intensifying production from existing stretches of cultivated land. Only in areas close to major demand centres, or enjoying good road or water links, were commercial farming and intensification practical possibilities. Ploughland increased more substantially than any other land-use realm in no fewer than 55 départements in the two decades up to mid-century, although the growing demand for wine enabled viticulture to perform that role in 10 départements in Aquitaine, Languedoc and Lyonnais, often advancing at the expense of arable farming (Fig. 10.4a). By contrast, the northern margin of vine growing was retreating significantly at this time. Important quantities of land were being converted to permanent grass to allow precocious pastoral specialisation in Normandy, Nivernais, the pays charentais and Pyrenees; while the process of afforestation was dominant in parts of Champagne, the middle Loire and the Alps. Large amounts of wasteland were converted to other uses in départements throughout southern, middle and western France; and substantial expanses

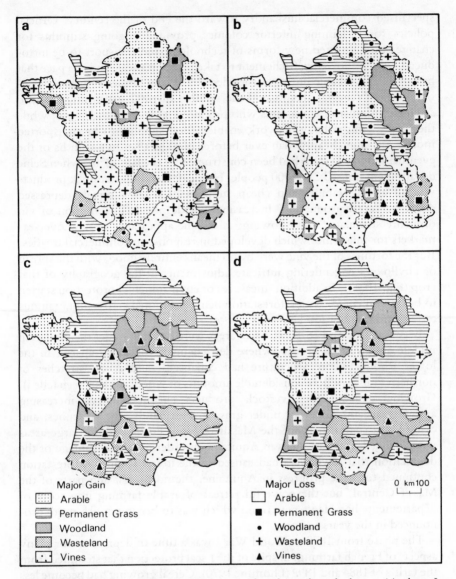

**Figure 10.4** Major changes in land-use realms by cadastral phases: (a) phase I; (b) phase II; (c) phase III; (d) ancien cadastre–1907.

of woodland were felled in the north and east of the country and along the southern fringe of the Massif Central.

The Second Empire and the early years of the Third Republic represented a complicated phase of transition from the earlier obsession with cereal production to the subsequent depression of grain growing and advance of specialised commercial husbandry toward the end of the century. Official policies for reclaiming interior colonies provided ongoing stimulus for change and permitted new forms of technology and transport to be introduced, which allowed défrichement to take place on a scale and at a pace that had been unthinkable during earlier decades. Programmes of urban development and renewal accentuated the volume of rural–urban migration to swell the number of consumers who produced no food of their own, while the newly built railway network enabled food supplies to be transported more easily and cheaply than ever before. No less than five-eighths of the eventual railway system had been constructed by the late 1870s, when Seine département housed 2 500 000 people. Home demand for livestock products was advancing rapidly and cheap supplies of foreign cereals depressed domestic cereal production. Increasingly widespread consumption of vin ordinaire stimulated vine growing, as did the appearance of new overseas markets for fine wines which developed in response to commercial treaties. But the fortunes of the vine were by no means entirely good, with the spread of phylloxera engendering intricate adjustments in the geography of that crop (Fig. 10.4b). In addition, measures of environmental protection started to be implemented, with afforestation being promoted across mountainous terrain and on poor lowland soils.

Between 1851 and 1879 extensive areas were brought under the plough in middle and western France (where défrichement continued apace), in the lower Rhône (where viticulture had collapsed), and across stretches of northern France (where considerable amounts of woodland had been felled). Growing demand for livestock products encouraged ever-increasing amounts of land to be put under grass in Normandy, the Ardennes, and eastern and central parts of the Massif Central. Further south, large areas were being devoted to vines in Aquitaine in order to take advantage of the devastation that phylloxera had caused in the lower Rhône. Afforestation continued to make progress in Aquitaine, the north-east and parts of the Massif Central, but the powerful retreat of arable farming in a score of départements heralded a new trend which was to become much more pronounced in the years to come.

The phase from 1879 to World War I was a time of depression for many aspects of French farming, in spite of the fiscal protection that stemmed from the tariffs of 1884 and 1892 (Lhomme 1970). Cereal growing had become less profitable and the hunger for arable land was well and truly over. In fact, the retreat of ploughland formed the major loss in no fewer than 51 départements (Fig. 10.4c). Only in 10 départements did it form the major element to advance at this time, as wasteland was still being cleared in Brittany and middle France, and viticulture was retreating in the Loire valley. At the same time, ploughland was being replaced by vines in Languedoc, but much more usual were the tendencies for arable land to be afforested, abandoned to waste or converted to permanent grass, with this latter trend taking place across much of northern France. The new 'industrial' vineyards of Languedoc came into full production soon after 1900 and many pays viticoles elsewhere in France, which had weathered the ravages of phylloxera, proved quite unable to cope with a flood of vin ordinaire from the Mediterranean south and Algeria. As a result, extensive areas of vines were lost in the south-west and the middle Loire, with only vineyards that produced fine wines being able to survive the aggressive competition of the Midi. By the eve of World War I, many other elements of the traditional economy had also declined substantially, with walnut trees and groves of chestnuts and olives occupying much less land than 100 years before, and several commercial crops like flax, hemp and madder having paled into insignificance because of competition from other commodities.

The fear of famine that had haunted generations of peasants and administrators disappeared gradually between 1815 and 1914. From being the pivot of the farming economy, around which all other activities were articulated and to which they were subordinate, grain growing had become just one form of agricultural specialisation among many others. By the turn of the century, cereal farming had acquired this status across much of the northern half of France, but by contrast many traditional aspects of agricultural activity still survived in less accessible pays that were much less open to innovation than areas traversed by efficient transport systems. The nation's arable realm declined by 1 429 040 ha from the late 1830s to the eve of World War I, with no fewer than 42 départements throughout eastern France, Normandy, the northern Paris Basin and Languedoc recording the retreat of arable as the most extensive decline in land use during that long period (Fig. 10.4d). In complete contrast, the advance of ploughland formed the most important increase in land use in 16 départements arching through Brittany and middle France, where large reserves of heath and rough grazing had been carved into new fields that, rather surprisingly, remained under the plough in the early 20th century. No less than 1 141 925 ha of wasteland disappeared, and this trend dominated negative changes in land use in 21 départements in Aquitaine as well as Armorica and middle France. Nonetheless, land abandonment was widespread and exceeded all other negative aspects of land-use change in 13 départements in the Pyrenees, southern Massif Central, Alps and eastern parts of the Paris Basin. Increasing emphasis on pastoral activities, especially cattle raising, was demonstrated by the fact that 2 051 825 ha were converted to permanent grass from other land uses between the 1830s and the early 20th century, and this was the most important positive trend in 40 départements in Vendée, Normandy and much of northern, eastern and central France (Martin 1966). Viticulture retreated by 594 835 ha nationwide and this negative trend was dominant in 13 départements in the pays charentais, Gascony and the southern Paris Basin. Only in Aude and Hérault did vine growing expand sufficiently to outweigh other advancing forms of land use. In addition, France contained

542 580 ha more cadastral woodland in 1908 than in the 1830s, with afforestation being the dominant process of land-use increase in 10 départements in Aquitaine, Champagne and Provence.

Despite all these complex changes, half of the land of France remained under the plough in the early 20th century, with 60 per cent being exceeded in a score of départements in a northern sweep from Pas-de-Calais through the Ile-de-France to Bourbonnais, and in a western band from northern Brittany to Poitou (Fig. 10.3b). This pattern reflected the fundamental resource base of the great plains and plateaux of the Paris Basin but also derived from the impact of widespread défrichement in middle France and the fact that other forms of land-use specialisation had developed in Normandy, eastern France and the Midi. By the early 1900s, the Massif Central and Normandy had become France's leading foci of pastoral activity, with 12 départements having over one-fifth of their land under permanent grass and 25 per cent being exceeded in seven of these (Fig. 10.3c). Vineyards occupied a mere 2.9 per cent of the whole country but they covered 18.5 per cent of Aude and 30.2 per cent of Hérault, with surviving vineyards in the far south, Bordelais and the pays of the Saône emerging quite sharply from Figure 10.3d. Extensive afforestation had transformed many landscapes, so that woodland occupied one-sixth of the land of France and exceeded 25 per cent in the north-east, Aquitaine and parts of the south (Fig. 10.3e). Just over a quarter of the country remained uncultivated and large amounts of wasteland were found in the far west of Brittany and across the southern third of France (Fig. 10.3f).

As well as these important changes between realms, profound modifications occurred within land-use realms during the long span of years from the July Monarchy to World War I. For example, wheat, oats and fodder crops advanced to occupy areas of ploughland that had formerly borne secondary crops or large shares of fallow. Modified versions of the classic biennial rotation with extensive fallows were widespread to the south of the Loire during the early 1870s, although by that time triennial rotations formed the second most common system in much of the Midi (Fig. 10.5b). Various forms of triennial rotation were dominant further north, with second place tending to be occupied by biennial systems in the north-east and quadriennial shifts in the north-west. The pattern must have changed to some extent in the decades that followed, but no subsequent information was gathered prior to World War I. Wheat cultivation had made substantial advances during the 19th century and in 1912 occupied more than 35 per cent of the arable realm in Aquitaine and the pays of the Saône and 30–35 per cent in Brie, eastern Armorica and Gascony (Fig. 10.5c). Despite increased cultivation in many regions, oats remained an essentially northern crop which covered one-fifth of ploughlands throughout the Paris Basin and exceeded 25 per cent in the Ile-de-France, Lorraine and Picardy (Fig. 10.5d). The vast traditional ségala of the Massif Central had contracted markedly by

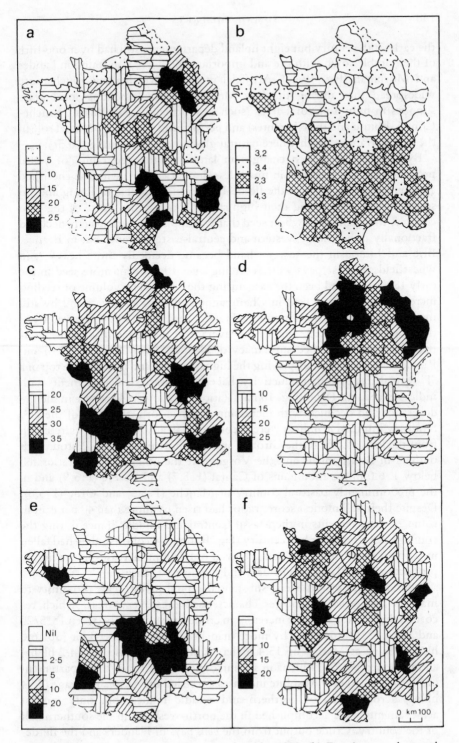

**Figure 10.5** (a) Fallow as percentage of arable, 1892. (b) Dominant and second-ranking rotations, 1873. (c) Wheat, (d) oats, (e) rye and (f) artificial meadows as percentage of arable, 1912.

the early 20th century but eight upland départements still had over one-fifth of their arable sown with rye and important foci were also found in Landes and Morbihan (Fig. 10.5e). Artificial meadows had become much more widespread than during the July Monarchy, supplementing local grass supplies in humid pastoral areas (such as Normandy, Nivernais, Franche-Comté, Poitou and the Charentes) and performing an even more vital role in the Mediterranean south where natural grassland was rare (Fig. 10.5f).

By contrast with such changes in land use, the geography of cereal productivity was modified in much less dramatic ways, with the cherished and long-fertilised pays of the Ile-de-France and Flanders retaining and, in some respects, enhancing their high status. During the course of the 19th century, the quantity of wheat seed that was applied to each hectare declined fractionally in northern, western and central-eastern France but in Beauce, Brie and Limousin the sower's hand became decidedly more heavy, and wheatfields in these pays were receiving over 10 per cent more seed in the early 1890s than had been the case during the 1830s. The volume of seeding increased by 5–10 per cent in Champagne and the south-west and by still more modest proportions in middle France. Seed : yield ratios, derived from the crop returns for 1892 and for the 'average year' cited in that source, confirm the long established supremacy of Flanders, with Nord (1 : 13.6) and Pas-de-Calais (1 : 11.4) displaying the highest values and forming the top of a 'T' shaped area of high productivity that encompassed six départements with indices exceeding 1 : 9.0 (Fig. 10.6a). Ratios were lower elsewhere in the Ile-de-France and declined with increasing distance form the stem of the 'T'. Similarly impressive indices were encountered in a 'V' of territory stretching from the lower Loire to the Gironde and then northwards into Poitou, with outliers in Bourbonnais and the Vosges. At the other extreme, ratios fell below 1 : 6.0 in the mountains of Cantal (1 : 5.7) and Lozère (1 : 5.9) and in the predominantly pastoral country of Manche (1 : 5.5) and Orne (1 : 5.8). Despite their still modest score, ratios had risen by more than 40 per cent in some 20 départements in the Massif Central and middle France during the course of the preceding half century (Fig. 10.6b). By contrast, they had fallen in parts of Normandy and had achieved only slight increases in the highly productive wheatlands of Flanders and the Ile-de-France.

Abundant use of fertilisers, substantial capital investment, proximity to markets and other advantages characterised these very areas, which recorded the highest absolute increases in crude wheat yields between 1815/20 and 1902/12. Output rose by more than 14 hl/ha in Eure-et-Loir (up 16.0 hl/ha), Seine-et-Marne (up 15.4 hl/ha), and Seine-et-Oise (up 14.2 hl/ha) and increased by 12–14 hl/ha in six surrounding areas plus the département of Nord (Fig. 10.6c). Absolute increases exceeded the national average (7.94 hl/ha) across much of northern and middle France, but only modest achievements were accomplished in the north-west and in the southern half of the country. Crude output from the *bons pays* of Flanders and the Ile-de-

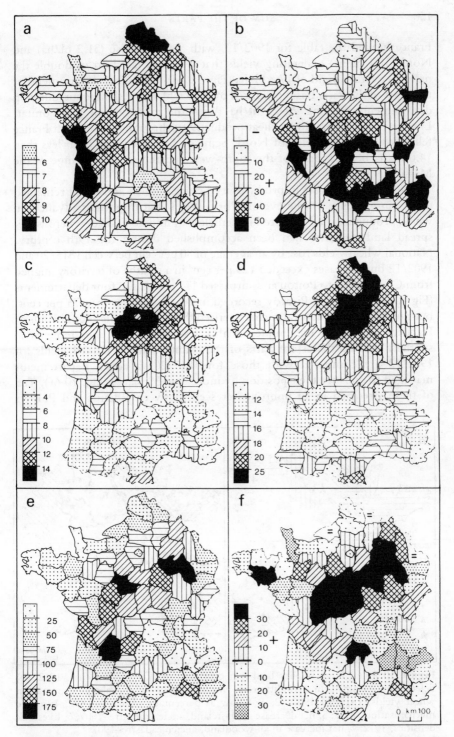

**Figure 10.6**  Wheat: (a) seed:yield ratios, 1892; (b) change in seed:yield ratios, 1837/52–1892 (per cent); (c) absolute increase in yields, 1815/20–1902/12 (hl/ha); (d) average yields, 1902/12 (hl/ha); (e) increase in yields, 1815/20–1902/12 (per cent); (f) yields, change in ranked positions, 1815/20–1902/12.

France headed the table for 1902/12, with Seine-et-Oise (31.2 hl/ha) and Nord (31.1 hl/ha) generating yields that were not far short of double the national average yield of 17.8 hl/ha (Fig. 10.6d). Five other départements at the heart of the Paris Basin produced over 25.0 hl/ha, with nine surrounding départements yielding 20–25 hl/ha. With the exception of the Comtat, Limagnes and Roussillon, wheat productivity throughout southern France fell below half the yields of Nord, Seine or Seine-et-Oise, with less than 14.0 hl/ha being typical of the south-west and the southern fringe of the Massif Central.

Traditional contrasts in productivity between north and south remained clearly in evidence on the eve of World War I, but important advances in wheat production had occurred nonetheless in middle France, where widespread land clearance had been accomplished during the 19th century. National wheat yields rose by an average of 80 per cent between 1815/20 and 1902/12 but increases exceeded 125 per cent in a swathe of territory aligned from Champagne to Poitou and surpassed 175 per cent in four départements (Fig. 10.6e). Brie and Picardy recorded increases of more than 75 per cent, but much more modest rates characterised the highly fertile ploughlands of Flanders as well as the unquestionably less productive départements in southern and northwestern parts of the country. When ranked yields for 1902/12 are compared with those for 1815/20, a dozen départements managed to enhance their position by more than 30 places (Fig. 10.6f). Ten of these stretched in an unbroken crescent between Rheims and Poitiers

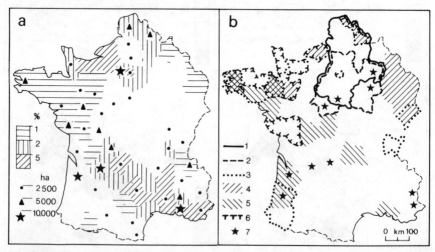

**Figure 10.7** (a) High quality land (per cent) and market gardens (ha), 1912. (b) Selected indicators of rural conditions: 1, wheat yields 1902/12; 2, absolute increase in wheat yields, 1815/20–1902/12; 3, fallow as percentage of arable 1892; 4, arable mechanisation, 1892; 5, net decrease in wasteland, ancien cadastre–1907; 6, livestock densities, 1912; 7, net increase in all woodland, ancien cadastre–1907.

and were surrounded by a cluster of départements, embracing Beauce, Burgundy and western Lorraine, which increased their position by more than 20 places. Many départements in the south and west of France descended in rank to compensate for such dramatic change in middle France, while only slight changes in position were possible for the highly productive départements toward the top of the league. These impressive changes of rank in middle France were made possible by decades of land improvement, which served to extend the traditional heartland of wheat growing in the Ile-de-France and the northern Paris Basin into a range of pays located immediately to the east and south. These same areas enjoyed the appropriate combination of physical, tenurial and financial conditions to achieve important relative and absolute increases in output. In many other parts of France the trend of specialisation had turned toward other activities, including pastoral farming in Armorica and the Massif Central, 'industrial' viticulture in Languedoc, and fruit and vegetable production in Provence, the pays of the Garonne and the environs of Paris and many other large cities (Fig. 10.7a).

## A final view

Seven indicators of rural conditions have been synthesised to produce Figure 10.7b which identifies départements that formed the top octile for each variable. As we have seen, crude output of wheat (hl/ha) was highest in northern and central parts of the Paris Basin in the early 1900s, although a slightly more southerly sweep of départements (plus Flanders) had achieved greatest absolute increases in crude output since 1815/20. By contrast, fallowing survived in the Paris Basin during the 1890s but had virtually disappeared in parts of Armorica, Normandy, the south-west and centre-east, as well as the départements of Nord and Seine. At that time, arable mechanisation was most pronounced in northeastern France, Upper Normandy, Flanders and Upper Brittany. Middle France, Aquitaine and Armorica displayed the strongest retreat of wasteland; and those first two regions, together with the Alps and Champagne, also underwent important afforestation. Finally, livestock husbandry had become especially concentrated across much of northwestern France as well as in Seine and Nord.

The testimony of earlier chapters has shown that significant changes in land use and agricultural activity had taken place in southern France between 1815 and 1914, but with the exception of afforestation and commercial viticulture these came nowhere near the top of the national league. Admittedly, many forms of fruit growing were essentially southern phenomena but they involved only small areas of land (Fig. 10.7a). Défrichement was also well represented in some pays south of the Loire, although that process symbolised the old order of land management rather than the new and had in any case given way to land abandonment in many districts by the early 1900s. By contrast, intensification emerged as an essentially northern process which manifested itself in individual départements in numerous distinctive ways. Despite significant agricultural changes during the 19th century, Flanders managed to retain its old-established supremacy, with Nord figuring in the top octile for no less than five of the seven indicators shown on Figure 10.7b (Gomart 1860). Not surprisingly, such an intensively worked area contained precious little waste or newly planted woodland. By one of the ironies of history, the holocaust of World War I was to change all that, reducing its thriving towns and fertile ploughlands to utter desolation.

In contrast with the achievements of many northern départements, central and southern parts of France displayed fewer and certainly less impressive examples of modernisation or intensification in the agricultural realm. On the eve of World War I, the great majority of farmers, operating outside the Paris Basin or working beyond more distant foci of pastoralism, viticulture or fruit growing, were still only loosely adapted to meeting the fluctuating requirements of the market. Instead they concentrated on producing the bulk of what their families needed to survive, going to market to sell any surpluses they generated and to acquire commodities they could not produce for themselves. Remaining beyond the dual embrace of commercialisation and specialisation, many peasant families managed to weather the storm of late 19th century agricultural depression (Wright 1964). Their modest accomplishments conspired to mask the high yields generated on modernised holdings in the Paris Basin and the Midi and ensured that, despite tightly localised improvements, low average productivity kept France close to the bottom of the league of Western Europe's farming nations (Table 10.1).

**Table 10.1**   Western Europe: comparative national indicators, 1912/13.

| | Average yield (qx/ha) | | | | Land use (per cent)† | | |
| --- | --- | --- | --- | --- | --- | --- | --- |
| | wheat | rye | oats | barley | cultivated | uncultivated | woodland |
| Denmark | 33.5 | 18.4 | 17.9 | 23.2 | 75.7 | 15.7 | 8.6 |
| Belgium | 25.2 | 22.0 | 25.6 | 27.0 | 65.4 | 5.8 | 17.8 |
| Netherlands | 24.4 | 20.1 | 22.1 | 26.2 | 67.0 | 25.1 | 7.9 |
| Germany | 23.6 | 19.1 | 21.9 | 22.2 | 64.8 | 9.3 | 25.9 |
| Great Britain | 21.2 | n.d. | 16.6 | 18.5 | 56.9 | 22.3 | 4.9 |
| France | 13.8 | 11.1 | 12.9 | 14.5 | 66.6 | 13.7 | 19.7 |
| Italy | 12.2 | 11.4 | 12.5 | 9.4 | 72.6 | 18.5 | 15.9 |
| Spain | 7.8 | 9.8 | 6.7 | 9.1 | 39.6 | 39.4 | 21.0 |
| Western Europe★ | 13.5 | 18.1 | 19.2 | 16.6 | 59.7 | 19.3 | 18.5 |

Source: Ministère de l'Agriculture 1914. Statistique agricole annuelle 1913. Paris: Imprimerie Nationale.

★Calculated as mean of the sum of national averages.

†National definitions vary; totals do not necessarily reach 100 per cent.

Underlying the new geometry of communications in the age of rail, a growing centralisation of wealth and power in Paris, and profound modifications in land use and landscape, the survival of a large landowning peasantry provided cultural continuity with times past and thereby ensured that change in the countryside was slow in the extreme. The challenge of interpreting this condition, and of associating the intricacies of the land of France with the complexities of the people who farmed it, through a coherent set of theory and a convincing set of explanations, remains to be met.

## Note

[1] Statements on the supply of various types of foodstuff to Paris prior to the construction of the railway network are provided by Husson (1856) and Vidalenc (1952, 1970).

# Bibliography

Abbreviations of journal titles used are as follows:

| | | | |
|---|---|---|---|
| AA | Annales Agronomiques | JAP | Journal d'Agriculture Pratique |
| AAF | Annales de l'Agriculture Française | JEH | Journal of Economic History |
| AB | Annales de Bourgogne | JHG | Journal of Historical Geography |
| AESC | Annales: Economies, Sociétés, Civilisations | LC | Le Cultivateur |
| | | N | Norois |
| AG | Annales de Géographie | REP | Revue d'Economie Politique |
| AH | Agricultural History | RGA | Revue de Géographie Alpine |
| AHR | Agricultural History Review | RGE | Revue Géographique de l'Est |
| AN | Annales de Normandie | RGPSO | Revue Géographique des Pyrénées et du Sud-Ouest |
| AS | Annales de Statistique | | |
| BSLG | Bulletin de la Société Languedocienne de Géographie | RHES | Revue d'Histoire Economique et Sociale |
| EHR | Economic History Review | RHMC | Revue d'Histoire Moderne et Contemporaine |
| ER | Etudes Rurales | | |
| ERH | Etudes Rhodaniennes | TIBG | Transactions, Institute of British Geographers |
| ISEA | Institut des Sciences Economiques Appliquées | | |

Abeausy, M. 1839–40. De l'état de la culture maraîchère aux environs de Paris. *JAP* **3**, 452–8.

Ackerman, E. B. 1977. Alternatives to rural exodus; the development of the commune of Bonnières-sur-Seine in the nineteenth century. *Fr. Hist. Stud.* **10**, 126–48.

Ackerman, E. B. 1978. *Village on the Seine. Tradition and change in Bonnières, 1815–1914.* Ithaca: Cornell University Press.

Anon. 1802. Département de l'Orne. *AS* **2**, 499–505.

Anon. 1802. Département du Doubs. *AS* **2**, 68.

Anon. 1802. Département du Drôme. *AS* **2**, 389–414.

Anon. 1803. Département de l'Ille-et-Vilaine. *AS* **7**, 81–156.

Anon. 1823. Du département de la Gironde. *L'Ami des Champs: Journal d'Agriculture du Département de la Gironde*, 33–44.

Anon. 1839. Sur l'introduction des instruments perfectionnés. *Bulletin de la Société Agricole du Département du Lot* **4**, 76–8.

Anon. 1840–1. Des défrichements en Basse-Bretagne. *JAP* **4**, 109–12.

Anon. 1846. Chemins de fer français. *Annuaire du Département de la Nièvre*, 129–33.

Anon. 1867. Enquête sur la situation et les besoins de l'agriculture. *Bulletin de la Société d'Agriculture de l'Aveyron*, 1–78.

Anon. 1887. De l'élevage du cheval en France. *Bulletin de la Société des Agriculteurs de la Somme*, 142–4, 172–4.

Anon. 1894. Inondations. *Revue Agricole, Industrielle, Historique et Artistique de l'Arrondissement de Valenciennes* **44**, 220.

Arbellot, G. 1973. La grande mutation des routes de France au milieu du XVIIIᶜ siècle. *AESC* **28**, 765–91.

Argouarch, G. 1884. Culture de l'ajonc. *Bulletin de la Société d'Agriculture de l'Arrondissement de Lorient*, 7–12.

Armengaud, A. 1951. Les débuts de la dépopulation dans les campagnes toulousaines. *AESC* **6**, 172–8.

Arqué, P. 1935. Problèmes d'assainissement et de mise en valeur des Landes de Gascogne. *RGPSO* **6**, 5–25.

Augé-Laribé, M. 1925. *L'agriculture pendant la guerre*. Paris: Presses Universitaires de France.

Augé-Laribé, M. 1945. Les statistiques agricoles. *AG* **53–4**, 81–92.

Augé-Laribé, M. 1955. *La révolution agricole*. Paris: Albin.

Avannes, M. 1834. Notice historique sur le département de l'Eure. *Bulletin de l'Ancienne Société d'Agriculture de l'Eure* **2**, 103–34.

Azambre, G. 1929. L'industrie laitière en Thiérache et dans le Hainaut français. *AG* **38**, 561–76.

Barral, J. A. 1852. Engrais et amendements dans le nord de la France. *JAP* **5**, 443–7.

Barral, J. A. 1853. Des terres à drainer en France. *JAP* **7**, 5–17.

Barral, J. A. 1856. *Drainage des terres arables*. Paris.

Barral, P. 1962. *Le Département de l'Isère sous la Troisième République, 1870–1940*. Paris: Colin.

Barral, P. 1968. *Les agrariens français de Méline à Pisani*. Paris: Colin.

Barral, P. 1979. Le monde de la terre. In *Histoire économique et sociale de la France*, F. Braudel and E. Labrousse (eds), vol. 4(i), 349–95. Paris: Presses Universitaires de France.

Baud, P. 1932. *L'industrie chimique en France; étude historique et géographique*. Paris: Masson.

Baudrillart, M. 1831. Mémoire sur le déboisement des montagnes. *AAF* **8**, 65–78.

Beaudoin, A. 1891. Note sur l'irrigation en France et en Algérie. *AA* **17**, 241–72, 305–25.

Bénévent, E. 1938. La vieille économie provençale. *RGA* **26**, 531–69.

Bergeron, L. 1970. Problèmes économiques de la France napoléonienne. *RHMC* **17**, 469–505.

Bergeron, L. 1972. *L'épisode napoléonien*. Paris: Seuil.

Bernard, G. 1978. *Le vignoble charentais*. Bordeaux: CERVIN.

Bernard, P. 1953. *Economie et sociologie de la Seine-et-Marne 1850–1950*. Paris: Colin.

Bethemont, J. 1972. *Le thème de l'eau dans la vallée du Rhône*. Saint-Etienne.

Billé, R. 1930. Le chemin de fer de Paris à Bordeaux. *AG* **39**, 449–67.

Blanqui, A. 1846. *Du déboisement des montagnes*. Paris.

Bloch, M. 1931. *Caractères originaux de l'histoire rurale française*. Paris: Colin.

Block, M. 1850. Les animaux domestiques en 1812, 1829 et 1839. *JAP* **1**, 739–43.

Block, M. 1860. *Statistique de la France*, 2 vols. Paris.

Bobin, R. 1926. La gâtine de Parthenay. *AG* **35**, 405–12.

Boitel, A. 1881a. Prairies et irrigations des Vosges. *AA* **7**, 32–73.

Boitel, A. 1881b. Prairies naturelles du bassin de la Saône. *AA* **7**, 524–50.

Bosc, M. 1826. Notes sur deux modes de culture propres à augmenter les produits de la Champagne craïeuse. *AAF* **33**, 60–7.

Bottin, M. 1833. Statistique agricole, arrondissement de Pau, Basses-Pyrénées. *LC* **8**, 244–54.

Bouchard, L. 1834. Des chemins vicinaux et de l'influence des maires sur leur viabilité. *AAF* **14**, 41–4.

Bougeatre, E. 1971. *La vie rurale dans le Mantois et le Vexin au XIXᵉ siècle*. Meulan.

Boulmier, A. 1951. L'outillage des champs dans le département de Saône-et-Loire. *RGL* **26**, 1–32.

Bourgnet, M. N. 1976. Race et folklore: l'image officielle de la France en 1800. *AESC* **31**, 802–23.

Bouscasse, M. 1867. Enquête sur la situation et les besoins de l'agriculture. *Bulletin de la Société d'Agriculture de La Rochelle*, 3–98.

Bozon, P. 1961. *La vie rurale en Vivarais*. Clermont-Ferrand: Faculté des Lettres.

Braudel, F. 1951. La géographie face aux sciences humaines. *AESC* **6**, 485–92.

Brosselin, A. 1977. La forêt bourguignonne. *AB* **49**, 119–33.

Brown, J. C. 1876. *Reboisement in France*. London: King.

Brunet, P. 1954. L'évolution du vignoble des Corbières vers la qualité. *RGPSO* **25**, 340–3.

Brunet, P. 1960. *Structure agraire et économie rurale des plateaux tertiaires entre la Seine et l'Oise*. Caen: Caron.

Brunet, R. 1959. Le vignoble entre Tarn et Garonne. *RGPSO* **30**, 135–67.

Brunhes, J. 1900. L'homme et la terre cultivée: bilan d'un siècle. *Bulletin de la Société Neuchâteloise de Géographie* **12**, 219–60.

Burat, A. 1851. Note sur la division de la propriété foncière en France. *AA* **2**, 408–13.

Burguière, A. 1977. *Bretons de Plozévet*. Paris: Flammarion.

Caralp, R. 1951. L'évolution de l'exploitation ferroviaire en France. *AG* **60**, 321–36.

Caron, F. 1979. *An economic history of modern France*. London: Methuen.

Carrière, F. and P. Pinchemel 1963. *Le fait urbain en France*. Paris: Colin.

Castang, C. 1967. La politique de mise en culture des terres à la fin d'ancien régime. Unpublished thesis for Doctorat en Droit.

Castellan, G. 1960. Fourrages et bovins dans l'économie rurale de la Restauration; l'exemple du département du Rhône. *RHES* **38**, 77–97.

Castellan, G. 1962. Les céréales dans l'économie rurale de la Restauration; l'exemple du département du Rhône. *RHES* **40**, 175–99.

Cavaillès, H. 1933. Le problème de la circulation dans les Landes de Gascogne. *AG* **42**, 561–82.

Cavaillès, H. 1946. *La route française: son histoire, sa fonction*. Paris: Colin.

Chabot, G. 1945. *La Bourgogne*. Paris: Colin.

Chambrelent, M. 1887. *Les Landes de Gascogne*. Paris.

Chaptal, J. A. 1819. *De l'industrie française*, 2 vols. Paris.

Châtelain, A. 1949. Brie, terre de passage. *AESC* **4**, 159–66.

Châtelain, A. 1956. La lente progression de la faux. *AESC* **11**, 495–9.

Chaunu, P. 1972. Malthusianisme démographique et malthusianisme économique. *AESC* **27**, 1–19.

Chavard, A. 1900. Les irrigations et les associations syndicales. *AA* **26**, 332–44.

Chavard, A. 1937. *Monographie agricole du département du Cher*. Paris.

Chevalier, L. 1950. *La formation de la population parisienne au XIXᵉ siècle*. Paris: Presses Universitaires de France.

Chevalier, M. 1838. *Des intérêts materiels en France; travaux publics, routes, canaux, chemins de fer*. Paris.

Chevalier, M. 1842. *Cours d'économie politique*. Paris.

Chombart de Lauwe, J. 1946. *Bretagne et Pays de la Garonne; évolution agricole comparée depuis un siècle*. Paris: Presses Universitaires de France.

Chorley, G. P. H. 1981. The agricultural revolution in northern Europe, 1750–1880; nitrogen, legumes and crop productivity. *EHR* **34**, 71–93.

Clapham, J. H. 1921. *Economic development of France and Germany, 1815–1914*. Cambridge: Cambridge University Press.

Claval, P. 1978. *Haute Bourgogne et Franche-Comté*. Paris: Flammarion.

Clout, H. D. 1973–4. Reclamation of wasteland in Brittany, 1750–1900. *Bull. Soc. Royale Géog. d'Anvers* **84**, 29–60.

Clout, H. D. 1977a. Les défrichements en Belgique au XIXᵉ siècle. *N* **24**, 101–109.

Clout, H. D. (ed.) 1977b. *Themes in the historical geography of France*. London: Academic Press.

Clout, H. D. 1979. Land-use change in Finistère during the eighteenth and nineteenth centuries. *ER* **73**, 69–96.

Clout, H. D. 1980. *Agriculture in France on the eve of the railway age*. London: Croom Helm.

Clout, H. D. and A. D. M. Phillips 1972. Fertilisants minéraux en France au XIXᵉ siècle. *ER* **45**, 9–28.

Clout, H. D. and A. D. M. Phillips 1973. Sugar beet production in the Nord département of France during the nineteenth century. *Erdkunde* **27**, 105–19.

Clout, H. D. and K. Sutton 1969. The cadastre as a source for French rural studies. *AH* **43**, 215–23.

Colin, G. 1973. Vignoble et vin de Champagne. *Trav. Inst. Géog. Reims* **15**, 1–92.

Collins, E. J. T. 1972–5. The diffusion of the threshing machine in Britain, 1790–1880. *Tools and Tillage* **2**, 16–33.

Colman, H. 1848. *The agriculture and rural economy of France, Belgium, Holland and Switzerland.* London: Petheram.

Corbin, A. 1975. *Archaïsme et modernité en Limousin au XIX$^e$ siècle, 1845–1880*, 2 vols. Paris: Rivière.

Cotheret, M. 1836. Considérations générales sur les forêts. *Travaux de la Société d'Emulation du Jura*, 26–35.

Coyaud, L. M. 1974. Richesse agricole et urbanisation. *RGE* **14**, 183–214.

Crubellier, M. 1948. Le Briançonnais à la fin de l'ancien régime. *RGA* **36**, 259–99, 335–71.

Crubellier, M. (ed.) 1975. *Histoire de la Champagne.* Toulouse: Privat.


Daudin, H. 1834. Exposé de l'état de l'industrie agricole dans le département de l'Oise. *Bulletin de la Société Agricole et Industrielle de l'Oise* **1**, 22–85.

De Buffon, N. 1852. Des marais de la France. *JAP* **4**, 5–10.

De Chabriol, D. 1822. Le repeuplement des bois dans le département du Cantal. *AAF* **17**, 377–81.

De Chambray, M. 1834. De l'agriculture et de l'industrie dans le province de Nivernais. *AAF* **13**, 4–22.

De Gasparin, M. 1846–7. Considerations sur les subsistences. *JAP* **4**, 353–62.

De Gasparin, M. 1851. Note sur les terrains du delta du Rhône. *AA* **1**, 488–96.

Delamarre, M. J. B. and H. Hairy 1971. *Techniques de production: l'agriculture.* Paris: Musées Nationaux.

De La Véronne, C. 1971. *La Brenne: histoire et traditions.* Tours.

De Magneville, M. 1827. Mémoire sur l'agriculture du département du Calvados. *Mémoires de la Société d'Agriculture et de Commerce de Caen* **2**, 41–52.

Demangeon, A. 1946. *La France: économique et humaine*, 2 vols. Paris: Colin.

Demangeon, S. 1928. L'approvisionnement de Paris en fruits et legumes. *AG* **37**, 97–121.

Demonet, M., P. Dumont and E. Le Roy Ladurie 1976. Anthropologie de la jeunesse masculine en France au niveau d'une cartographie cantonale, 1819–30. *AESC* **31**, 700–60.

De Penanster, H. 1887. Inauguration du monument élevé à la mémoire de M. Jules Rieffel. *Bulletin de l'Association Bretonne*, 30–4.

De Quincy, C. 1831. Situation agricole du département de l'Indre. *LC* **5**, 8–13.

De Reparaz, G. A. 1966. *Le Plateau de Saint-Christol. Etude de géographie rurale en Haute Provence.* Aix-en-Provence: Faculté des Lettres.

Derruau, M. 1949. *La Grande Limagne auvergnate et bourbonnaise.* Clermont-Ferrand: Delaunay.

De Sainte-Colombe, M. 1828. Notice sur l'instruction publique, l'agriculture et l'industrie de l'arrondissement d'Yssingeaux. *Annales de la Société d'Agriculture, Sciences, Arts et Commerce du Puy* **3**, 80–151.

Désert, G. (ed.) 1978. *La Normandie de 1900 à nos jours.* Toulouse: Privat.

Dessalles, P. 1937. *Monographie agricole du département de la Haute-Vienne.* Limoges.

De Tocqueville, A. 1843–4. Des principales questions agricoles; bestiaux, vins, laines. *JAP* **1**, 97–107.

Devailly, G. (ed.) 1980. *Histoire du Berry.* Toulouse: Privat.

Dezeimeris, M. 1841–2. Des moyens d'améliorer l'agriculture en France. *JAP* **5**, 1841–2, 568–76.

Dion, J. 1970. Les forêts de la France du nord-est. *RGE* **10**, 155–277.

Dion, R. 1934. *Essai sur la formation du paysage rural français.* Tours: Arrault.

Dion, R. 1959. *Histoire de la vigne et du vin en France.* Paris.

Dion, R. 1961. Le bon et beau pays nommé Champagne pouilleuse. *Inform. Géog.* **25**, 209–14.

Doé, M. 1832. Sur l'état présent de la culture en Sologne. *AAF* **10**, 339–44.

Douguedroit, A. 1976. *Les Paysages forestiers de Haute-Provence et des Alpes-Maritimes.* Aix-en-Provence: Edisud.

Drouyn de Lhuys, M. 1875. Le moissonnage mécanique dans le département de Seine-et-Marne en 1874. *Bulletin de la Société d'Agriculture de Melun*, 34–7.

Dubois, L. 1827. De l'engraissement des boeufs dans la vallée d'Auge. *AAF* **39**, 169–80.

Dubost, P. C. 1876. Essai de statistique de la population agricole en France. *AA* **2**, 542–74.

Dubuc, R. 1938. L'approvisionnement de Paris en lait. *AG* **47**, 257–66.

Duby, G. and A. Wallon (eds) 1976. *Histoire de la France rurale de 1789 à 1914*. Paris: Seuil.

Duffoure-Bazin, P. 1840. Agriculture; ce qu'elle est dans le midi, le centre et le nord de la France. *LC* **12**, 577–93.

Dufour, J. 1981. *Agriculture et agriculteurs dans les campagnes mancelles*. Le Mans.

Dugied, P. H. 1819. *Projet de boisement des Basses-Alpes*. Paris.

Dulac, A. 1900. Commerce des produits agricoles; le bétail et la viande de boucherie. *AA* **26**, 430–72.

Dumas, E. 1851. Note sur les mesures législatives prises en faveur du drainage en Angleterre. *AA* **2**, 383–9.

Dupâquier, J. 1972. La non-révolution du XVIIIᶜ siècle. *AESC* **27**, 80–4.

Dupeux, G. 1962. *Aspects de l'histoire sociale et politique du Loir-et-Cher 1848–1914*. Paris: Mouton.

Dupeux, G. 1974. La croissance urbaine en France au XIXᶜ siècle. *RHES* **52**, 173–9.

Dupont, M. 1847. Rapport sur les desséchements opérés en Queyries. *Annales de la Société d'Agriculture de la Gironde* **2**, 204–22.

Estier, R. 1976. Le temps des dépressions. In *Histoire des paysans français du XVIIIᶜ siècle à nos jours*, J. P. Houssel (ed.), 299–400. Roanne: Horvath.

Evrard, F. 1923. Les grandes fermes entre Paris et la Beauce. *AG* **32**, 210–26.

Fabre, J. M. 1829. Mémoire pour servir à la statistique du département du Cher. *Bulletin de la Société d'Agriculture du Département du Cher* **2**, 409–80.

Faucher, D. 1934. Polyculture ancienne et assolement biennal dans la France méridionale. *RGPSO* **5**, 241–55.

Faucher, D. 1954. *Le paysan et la machine*. Paris: Minuit.

Faucher, D. 1961. L'assolement triennal en France. *ER* **1**, 7–17.

Fayolle, G. 1977. *La vie quotidienne en Périgord au temps de Jacquou le Croquant*. Paris: Hachette.

Fénelon, P. 1978. *Les pays de la Loire*. Paris: Flammarion.

Festy, O. 1956. Les enquêtes agricoles en France de 1800 à 1815. *RHES* **34**, 43–59.

Festy, O. 1957. Le progrès de l'agriculture française durant le Premier Empire. *RHES* **35**, 266–92.

Fiétier, R. (ed.) 1977. *Histoire de la Franche-Comté*. Toulouse: Privat.

Fléchet, J. P. 1967. L'évolution agricole de la Dombes. *RGL* **42**, 39–79.

Fohlen, C. 1973. France, 1700–1914. In *Fontana economic history of Europe. The emergence of industrial societies*, C. M. Cipolla (ed.), 7–75. London: Fontana.

Fossier, R. (ed.) 1974. *Histoire de la Picardie*. Toulouse: Privat.

Fox, E. W. 1971. *History in geographic perspective*. New York: Norton.

Galtier, G. 1968. La reconversion du vignoble du Languedoc-méditerranéen. *BSLG* **2**, 155–62.

Gandilhon, R. 1969. La série M (administration générale) des archives départementales. *Revue d'Histoire* **241**, 147–62.

Garnier, B. 1975. Pays herbagers, pays céréaliers et pays ouverts en Normandie, XVIᶜ – début du XIXᶜ siècle. *RHES* **53**, 493–525.

Garrier, G. 1967. Les enquêtes agricoles du XIXᶜ siècle; une source contestée. *Cahiers d'Histoire*, 105–13.

Garrier, G. 1973. *Paysans du Beaujolais et du Lyonnais, 1800–1970*, 2 vols. Grenoble: Presses Universitaires de Grenoble.

Garrier, G. 1974. Aspects et limites de la crise phylloxérique en Beaujolais. *RHES* **52**, 190–208.

Garrier, G. 1977. Les enquêtes agricoles décennales du XIXᶜ siècle; essai d'analyse critique. In *Pour une histoire de la statistique*, Institut de la Statistique et des Etudes Economiques, 269–79. Paris: INSEE.

Gaussen, H. 1932. L'oeuvre des forestiers aux Pyrénées françaises. *RGPSO* **3**, 385–414.

Gay, F. 1967. *La Champagne du Berry*. Bourges: Tardy.

George, P. 1935. *La région du Bas-Rhône*. Paris: Ballière.

George, P. and C. Hughes 1933. Comment transformera-t-on la Camargue? *ERH* **9**, 26–68.

Gillardot, P. 1972. Forêts et landes de Sologne. *N* **19**, 641–72.

Gille, B. 1964. *Les sources statistiques de l'histoire de la France des enquêtes du XVII$^e$ siècle à 1870.* Paris: Minard.

Girard, A. C. 1901. Recherches sur l'utilisation de l'ajonc. *AA* **27**, 5–44.

Girard, L. 1952. *La politique des travaux publics du Second Empire*. Paris: Colin.

Gobin, A. 1859. Etat présent de l'agriculture en France. *AAF* **13**, 250–7.

Gomart, C. 1860. La culture flamande. *AAF* **16**, 530–9.

Gonnet, P. 1955. Esquisse de la crise économique en France de 1827 à 1832. *RHES* **33**, 249–91.

Goubert, P. 1957. Les techniques agricoles dans les pays picards. *RHES* **35**, 24–40.

Goujon, P. 1976. Le temps des révolutions inachevées. In *Histoire des paysans français du XVIII$^e$ siècle à nos jours*, J. P. Houssel (ed.), 123–298. Roanne: Horvath.

Goy, J. and E. Le Roy Ladurie 1982. *Tithe and agrarian history from the fourteenth to the nineteenth centuries*. Cambridge: Cambridge University Press.

Grandeau, L. 1885. *La production agricole en France*. Paris.

*Grande encyclopédie*, n.d. Paris.

Grantham, G. W. 1975. Scale and organisation in French farming, 1840–1880. In *European peasants and their markets*, W. N. Parker and E. L. Jones (eds), 293–326. Princeton: Princeton University Press.

Grantham, G. W. 1978. The diffusion of the new husbandry in northern France, 1815–1840. *JEH* **38**, 311–37.

Gras, C. 1979. Géographies régionales des années 1900. In *Histoire économique et sociale de la France*, F. Braudel and E. Labrousse (eds), vol. 4 (i), 323–47. Paris: Presses Universitaires de France.

Guellec, A. 1979. *Département et unité rurale; l'exemple des Côtes-du-Nord*. Poitiers: Norois.

Guillard, J. (ed.) 1980. *Des arbres et des hommes*. Le Paradou: Actes Sud.

Guillon, J. M. 1905. *Etude générale de la vigne*. Paris: Masson.

Guyetan, S. 1856. Situation agricole de la Dombes, Ain. *AAF* **7**, 225–7.

Guyot, J. 1868. *Etudes des vignobles de France*. Paris: Imprimerie Impériale.


Hau, M. 1976. *La croissance économique de la Champagne*. Paris: Ophrys.

Haudricourt, A. G. and M. J. B. Delamarre 1955. *L'homme et la charrue à travers le monde*. Paris: Gallimard.

Henderson, W. O. 1967. *The industrial revolution on the Continent*. London: Cass.

Henry, S. 1943. *La Forêt de Bouconne*. Toulouse: Privat.

Heuzé, G. 1852. Machine à battre mue par la vapeur. *JAP* **5**, 64–9.

Heuzé, G. 1868. *La France du sud-ouest*. Paris.

Higounet, C. (ed.) 1978. *Recherches sur l'histoire de l'occupation du sol du Périgord*. Paris: Centre National de la Recherche Scientifique.

Hitier, H. 1899. La statistique agricole de la France. *AG* **8**, 350–7.

Hitier, J. 1901. L'agriculture moderne et sa tendance à s'industrialiser. *REP* **15**, 105–17, 329–66, 630–73, 752–74.

Hitier, J. 1902. La transformation de l'outillage agricole et l'agriculture moderne. *REP* **16**, 753–76.

Hohenberg, P. 1972. Change in rural France in the period of industrialisation 1830–1914. *JEH* **32**, 219–40.

Hoslin, M. 1850. Fabrication de la chaux en Basse-Bretagne. *JAP* **1**, 565–9.

Houssel, J. P. (ed.) 1976. *Histoire des paysans français du XVIII$^e$ siècle à nos jours*. Roanne: Horvath.

Hubscher, R. H. 1979–80. *L'agriculture et la société rurale dans le Pas-de-Calais du milieu du XIX<sup>e</sup> siècle à 1914*, 2 vols. Arras: Commission Départementale des Monuments Historiques du Pas-de-Calais.

Huetz de Lemps, A. (ed.) 1978. *Géographie historique des vignobles*, 2 vols. Paris: Centre Nationale de la Recherche Scientifique.

Huffel, G. 1904. *Economie forestière*, 3 vols. Paris: Laveur.

Husson, A. 1856. *Les consommations de Paris*. Paris.

Inspecteurs de l'Agriculture 1843. *Agriculture française; Haute-Garonne*. Paris.

Inspecteurs de l'Agriculture 1843. *Agriculture française; Hautes-Pyrénées*. Paris.

Inspecteurs de l'Agriculture 1844. *Agriculture française; Côtes-du-Nord*. Paris.

Inspecteurs de l'Agriculture 1845. *Agriculture française; Tarn*. Paris.

Inspecteurs de l'Agriculture 1847. *Agriculture française; Aude*. Paris.

Jardin, A. and A. J. Tudesq 1972. *La France des notables, 1815–48*. 2 vols. Paris: Seuil.

Jeantet-Maret, R. 1940–1. La banlieue maraîchère et le commerce des légumes à Lyon. *ERH* **16**, 221–76.

Joigneaux, P. 1847. Péregrinations agronomiques dans le département de la Côte-d'Or. *Revue Agricole et Industrielle de la Côte-d'Or* **1**, 379–89.

Jorré, G. 1971. *Le terrefort toulousain et lauragais*. Toulouse: Privat.

Jusseraud, F. 1841–2. Statistique agricole de la commune de Vensat. *Bulletin Agricole du Puy-de-Dôme* **1**, 1–92.

Klatzmann, J. 1955. *La localisation des cultures et des productions animales en France*. Paris: INSEE.

Klatzmann, J. 1961. Les limites du calcul économique en agriculture. *ER* **1**, 50–6.

Labande, E. R. (ed.) 1976. *Histoire du Poitou, du Limousin et des Pays Charentais*. Toulouse: Privat.

Labrousse, E. (ed.) 1956. *Aspects de la crise et de la dépression de l'économie française au milieu du XIX<sup>e</sup> siècle, 1846–51*. Paris: Centrale.

Labrousse, E., R. Romano and F. G. Dreyfus 1970. *Le prix du froment en France au temps de la monnaie stable 1726–1913*. Paris: SEVPEN.

Laffargue, M. 1839. Statistique agricole. *Bulletin de la Société Agricole et Industrielle du Département du Lot* **4**, 153–9.

Lamairesse, J. 1861. Moissonneuses dans la Marne. *AAF* **17**, 236–8.

Landureau, A. 1883. Etudes sur les causes de la diminution de la culture du lin en France. *AA* **9**, 289–96.

Larroquette, A. 1924. *Les Landes de Gascogne et la forêt landaise*. Mont-de-Marsan: Dupeyron.

Latreille, A. (ed.) 1975. *Histoire de Lyon et du Lyonnais*. Toulouse: Privat.

Laurent, R. 1976. Le secteur agricole. In *Histoire économique et sociale de la France*, F. Braudel and E. Labrousse (eds), vol. 3(ii), 619–767. Paris: Presses Universitaires de France.

Laurent, R. 1978. Les quatre âges du vignoble du Bas-Languedoc et du Roussillon. In *L'economie et société en Languedoc-Roussillon de 1789 à nos jours*, G. Cholvy (ed.), 11–44. Montpellier: Faculté des Lettres.

Lebeau, R. 1955. *La vie rurale dans les montagnes du Jura méridional*. Lyons: Patissier.

Le Bon, G. 1862. La Brenne; recherches sur la fièvre intermittente. *AAF* **19** 471–81, 521–7, **20** 37–43.

Le Boyer, J. 1832. *Notices sur le département de la Loire-Inférieure*. Paris.

Le Bras, H. and E. Todd 1981. *L'invention de la France*. Paris: Librairie Générale Française.

Lebrun, F. (ed.) 1972. *Histoire des Pays de la Loire*. Toulouse: Privat.

Lefebvre, G. 1962. *Etudes orléanaises*. Paris: Commission d'Histoire Economique et Sociale.

Le Gallais, A. 1869. La culture à vapeur en 1868. *AAF* **33**, 190–5.

Legoyt, A. 1863. De l'état actuel de l'agriculture dans quelques états de l'Europe. *Journal de la Société de Statistique de Paris*, 193.

Léonce de Lavergne, G. 1857. *L'agriculture et la population*. Paris: Guillaumin.

Léonce de Lavergne, G. 1861. *Economie rurale de la France depuis 1789*. Paris: Guillaumin.

Le Play, F. 1879. *Les ouvriers européens*, 6 vols. Paris: Larcher.

Lerolle, L. 1852. Culture de la betterave dans le Nord. *JAP* **15**, 402–406.

Leroy, M. 1829. L'avantage des desséchemens. *LC* **1**, 72–5.

Le Roy Ladurie, E. 1972. *Times of feast, times of famine: a history of climate since the year 1000*. London: George Allen & Unwin.

Levy-Leboyer, M. 1972. *Le revenu agricole et la rente foncière en Basse-Normandie. Etude de croissance régionale*. Paris: Klincksieck.

Lhomme, J. 1970. La crise agricole à la fin du XIX^e siècle en France. Essai d'interpretation économique et sociale. *Rev. Econ.* **21**, 521–53.

Livet, G. 1942. La Double. *RGPSO* **13**, 170–260.

Livet, R. 1962. *Habitat rural et structures agraires en Basse-Provence*. Aix-en-Provence: Ophrys.

Livet, R. 1978. *Provence, Côte-d'Azur et Corse*. Paris: Flammarion.

Lullin de Châteauvieux, F. 1837–8. Etat de l'agriculture en Europe: France. *JAP* **1** 305–12, 458–97.

Lullin de Châteauvieux, F. 1843. *Voyages agronomiques en France*, 2 vols. Paris.


Macdonald, S. 1975. The progress of the early threshing machine. *AHR* **23**, 63–77.

Malte-Brun, M. 1833. *Universal geography*, vol. 8. Boston.

Mangon, H. 1860. Création et exploitation d'un domaine de 500 hectares dans les Landes de Gascogne par M. Chambrelent. *AAF* **15**, 241–50.

Manry, A. G. (ed.) 1974. *Histoire de l'Auvergne*. Toulouse: Privat.

Marcilhacy, C. 1962. *Le Diocèse d'Orléans sous l'episcopat de Mgr Dupanloup, 1849–1878*. Paris: Plon.

Marot, R. 1958. *Pathologie régionale de la France*. Paris: Ministère de la Santé Publique.

Marres, P. 1935. L'évolution de la viticulture dans le Bas-Languedoc. *BSLG* **6**, 26–58.

Marres, P. 1942. Notes de géographie caussenarde. *AG* **51**, 175–86.

Martin, G. 1966. Evolution de l'agriculture en Auxois de 1840 à 1939. *Cah. Assoc. Univ. de l'Est* **11**.

Martin, G. and P. Martenot 1909. *La Côte-d'Or: étude d'économie rurale*. Dijon: Venot.

Maspetiol, R. 1946. *L'ordre éternel des champs*. Paris: Médicis.

Masurel, Y. 1958. L'influence des crises de l'oïdium et du phylloxéra sur l'évolution du vignoble provençal. *Acta Geog.* **28**, 14–22.

Maury, R. 1976. Les variations de l'espace boisé en Touraine. *Actes du 97^e Congrès National des Sociétés Savantes, Nantes, 1972*, 71–103.

May, M. G. 1930. Le chemin de fer de Paris à Marseille. *AG* **39**, 376–94.

Meinig, D. 1978–9. The continuous shaping of America. *Am. Hist. Rev.* **83**, 1186–217.

Mercadier, M. 1802. Description abrégée du département de l'Ariège. *AS* **4**, 5–128.

Mergoil, G. 1978. Une tentative de reconquête agro-pastorale; le débroussaillement des Causses de Quercy. In *Etudes géographiques offertes à Louis Papy*, 253–61. Bordeaux: Faculté des Lettres.

Merley, J. 1974. *La Haute-Loire de la fin de l'ancien régime aux débuts de la troisième république, 1776–1886*. Le Puy: Archives Départementales.

Meynier, A. 1931. *A travers le Massif Central. Ségalas, Levézou, Châtaigneraie*. Aurillac: Editions USHA.

Miège, J. 1961. *La vie rurale du sillon alpin: étude géographique*. Paris: Génin.

Ministère de l'Agriculture, du Commerce et des Travaux Publics 1865–6. *Commission des engrais*, 2 vols. Paris. Imprimerie Impériale.

Ministère de l'Agriculture 1937. *Monographie agricole du département de l'Isère*. Paris.

Ministère de l'Agriculture 1937. *Monographie agricole du Côte-d'Or*. Paris.

Ministère des Finances 1869. *Enquête sur les incendies de forêts dans la région des Maures et de l'Esterel.* Paris: Imprimerie Impériale.

Ministère des Finances 1873. *Enquête sur les incendies de forêts dans la région des Landes de Gascogne.* Paris: Imprimerie Nationale.

Moll, L. 1838–9. Etat de l'agriculture dans le département de l'Indre. *JAP* 2, 7–16.

Mollat, M. (ed.) 1971. *Histoire de l'Ile-de-France et de Paris.* Toulouse: Privat.

Monteil, A. A. 1803. *Description du département de l'Aveyron.* Paris.

Moreau, J. P. 1958. *La vie rurale dans le sud-est du bassin parisien.* Paris: Belles Lettres.

Morineau, M. 1976. The agricultural revolution in nineteenth-century France: comment. *JEH* 36, 436–7.

Mugriet, B. 1821. Aperçu sur la situation rurale du second arrondissement du département des Landes. *AAF* 16, 5–64.

Mulliez, J. 1979. Du blé, mal nécessaire; réflexions sur les progrès de l'agriculture de 1750 à 1850. *RHMC* 26, 3–47.

Musset, L. 1952. Observations sur l'ancien assolement biennal du Roumois et du Lieuvin. *AN* 12, 143–50.

Musset, R. 1908. La limite de la culture de la vigne dans l'ouest de la France. *AG* 17, 268–70.

Nanton, P. 1957–63. *Atlas linguistique et ethnographique du Massif Central*, 4 vols. Paris: Centre National de la Recherche Scientifique.

Napo, F. 1971. *1907: la révolte des vignerons.* Toulouse: Privat.

Naville, J. 1850. De l'assainissement des terres et du drainage. *JAP* 1, 63–9.

Newell, W. H. 1973. The agricultural revolution in nineteenth-century France. *JEH* 33, 697–731.

Nicod, J. 1956. Grandeur et décadence de l'oléoculture provençale. *RGA* 44, 247–95.

Noirot, M. 1838. Agriculture; sur son état actuel en France. *LC* 14, 129–35.

O'Brien, P. O. and C. Keyder 1978. *Economic growth in Britain and France, 1780–1914.* London: George Allen & Unwin.

Pacoud, M. 1805. Recherches sur les causes générales et particulières de l'insalubrité de la ci-devant Dombes. *Travaux de la Société d'Emulation et d'Agriculture de l'Ain*, 15–16.

Papy, L. 1947–8. L'ancienne vie pastorale dans la grande Lande. *RGPSO* 18–19, 5–16.

Papy, L. 1978. *Les Landes de Gascogne et la Côte d'Argent.* Toulouse: Privat.

Parisse, M. (ed.) 1978. *Histoire de la Lorraine.* Toulouse: Privat.

Passy, H. 1846. *Des systèmes de culture et de leur influence sur l'économie sociale.* Paris.

Pautard, J. 1965. *Les disparités régionales dans la croissance de l'agriculture française.* Paris: Gauthier-Villars.

Peeters, A. 1975. Les plantes tinctoriales dans l'économie du Vaucluse au XIXᵉ siècle. *ER* 60, 41–56.

Perpillou, A. 1927. L'évolution économique du Limousin méridional. *AG* 36, 509–27.

Perpillou, A. 1940. *Cartographie du paysage rural limousin.* Chartres: Durand.

Perpillou, A. 1977. *Utilisation agricole du sol en France; première moitié du XIXᵉ siècle.* Paris: CNRS.

Perpillou, A. 1977. *Utilisation agricole du sol en France; seconde moitié du XXᵉ siècle.* Paris: CNRS.

Perpillou, A. n.d. *Utilisation agricole du sol en France; première moitié du XXᵉ siècle.* Paris: CNRS.

Phillips, A. D. M. and H. D. Clout 1970. Underdrainage in France during the second half of the nineteenth century. *TIBG* 51, 71–94.

Picot de la Peyrouse, P. I. 1819. *The agriculture of a district in the South of France.* London: Harding.

Pitié, J. 1971. *Exode rural et migrations intérieures en France.* Poitiers: Norois.

Plaisse, A. 1963. Histoire agraire et prospective. *RHES* 41, 289–312.

Plessis, A. 1973. *De la fête impériale au mur des fédérés, 1852–1871.* Paris: Seuil.

Pouthas, C. H. 1956. *La population française pendant la première moitié du XIXᵉ siècle*. Paris: Presses Universitaires de France.

Price, R. 1975. The onset of labour shortage in nineteenth-century French agriculture. *EHR* **28**, 260–79.

Price, R. 1981. *An economic history of modern France, 1730–1914*. London: Macmillan.

Prince, H. C. 1977. Regional contrasts in agrarian structures. In *Themes in the historical geography of France*, H. D. Clout (ed.), 129–84. London: Academic Press.

Prothero, R. E. 1908. *The pleasant land of France*. London: Nelson.

Purvis, M. A. 1839–40. Notes d'un journal agronomique dans les Vosges. *JAP* **3**, 401–409.

Puvis, M. 1840–1. Agriculture de la Flandre française et de la Belgique. *JAP* **4**, 160–6.

Puvis, M. 1843–4. De la culture des jardins en France. *JAP* **1**, 410–19.

Puvis, M. 1845–6. Du climat de l'agriculture du sud-est de la France. *JAP* **3**, 193–206, 241–57.

Puvis, M. 1850. Emploi du noir animal dans les défrichements. *JAP* **1**, 204–209.

Reed, J. L. 1954. *Forests of France*. London: Faber.

Reinhard, M. 1923. Le pays d'Auge. *AG* **32**, 33–40.

Richard, J. (ed.) 1978. *Histoire de la Bourgogne*. Toulouse: Privat.

Richard, M. 1927. Les forêts du plateau de Langres. *RGA* **15**, 533–66.

Richardson, G. G. 1877. *The corn and cattle producing districts of France*. London: Cassell.

Risler, E. 1897. *Géologie agricole*, 4 vols. Paris.

Robert-Muller, C. 1932. Les machines agricoles en Bretagne. In *Melanges géographiques offerts à Raoul Blanchard*, 537–50. Grenoble.

Rollet, C. 1970. L'effet des crises économiques au début du XIXᵉ siècle sur la population. *RHMC* **17**, 391–410.

Rousselle, A. 1877. Les polders de la Baie du Mont-Saint-Michel. *AA* **3**, 429–33.

Ruttan, V. W. 1978. Structural retardation and the modernisation of French agriculture. *JEH* **38**, 714–28.

Salmon, M. 1840–1. Notice sur l'agriculture du canton de la Flèche. *Bulletin de la Société d'Agriculture des Sciences et Arts du Mans*, 41–8.

Sargent, F. O. 1958. The persistence of communal tenure in French agriculture. *AH* **32**, 100–108.

Sargent, F. O. 1961. Feudalism to family farms in France. *AH* **35**, 193–201.

Schnitzler, J. H. 1846. *Statistique générale méthodique et complète de la France*, 4 vols. Paris.

Sée, H. 1927. *La vie économique de la France sous la monarchie censitaire 1815–1848*. Paris: Alcan.

Sentou, J. 1947–8. Les facteurs de la révolution agricole dans le Narbonnais. *RGPSO* **18–19**, 89–104.

Sermet, J. 1930. L'aménagement des marais de l'ouest. *RGPSO* **1**, 235–41.

Séverin-Canal, M. 1934. Quelques aspects de l'économie agricole du Tarn-et-Garonne vers le milieu du XIXᵉ siècle. *RGPSO* **5**, 57–84.

Sigaut, F. 1976. Pour une cartographie des assolements en France au début du XIXᵉ siècle. *AESC* **31**, 631–43.

Sion, J. 1909. *Les paysans de la Normandie orientale*. Paris: Colin.

Soboul, A. 1968. Survivances féodales dans la société rurale au XIXᵉ siècle. *AESC* **23**, 965–86.

Solle, H. 1981. L'utilisation agricole du sol en France: les cartes Aimé Perpillou. *Acta Geog.* **45**, 1–25.

Sornay, J. 1934. Les forêts du département du Rhône. *ERH* **10**, 113–25.

Sorre, M. 1907. La plaine du Bas-Languedoc. *AG* **16**, 414–29.

Stevenson, W. I. 1978. La vigne américaine; son rayonnement et importance dans la viticulture hérautaise au XIXᵉ siècle. In *Economie et société en Languedoc-Roussillon de 1789 à nos jours*, G. Cholvy (ed.), 69–85. Montpellier: Faculté des Lettres.

Stevenson, W. I. 1980. The diffusion of disaster; the phylloxera outbreak in the département of the Hérault 1862–80. *JHG* **6**, 47–63.

Stevenson, W. I. 1981. *Viticulture and society in the Hérault (France) during the phylloxera crisis, 1862–1907.* Unpubl. PhD thesis, University of London.

Surell, A. 1870. *Etude sur les torrents des Hautes-Alpes,* 2nd edn. Paris.

Suret-Canale, J. 1958. L'état économique et social de la Mayenne au milieu du XIX<sup>e</sup> siècle. *RHES* **36**, 294–331.

Sutton, K. 1969. La triste Sologne; l'utilisation du sol dans une région française à l'abandon au début du XIX<sup>e</sup> siècle. *N* **16**, 7–30.

Sutton, K. 1971. The reduction of wasteland in the Sologne; nineteenth-century French regional improvement. *TIBG* **52**, 129–44.

Sutton, K. 1973. A French agricultural canal; the canal de la Sauldre. *AHR* **21**, 51–6.

Sutton, K. 1977. Reclamation of wasteland. In *Themes in the historical geography of France,* H. D. Clout (ed.), 247–300. London: Academic Press.

Taillefer, F. (ed.) 1978. *Histoire du Rouergue.* Toulouse: Privat.

Tarot, J. B. 1840. Population de la Bretagne. *Agriculture de l'Ouest* **1**, 539–50.

Tessier, M. 1818. Notes recueillies à Perpignan sur des améliorations proposées et entreprises dans le pays. *AAF* **3**, 176–88.

Thuillier, G. 1956. Les transformations agricoles en Nivernais de 1815 à 1840. *RHES* **34**, 426–56.

Toutain, J. C. 1961. Le produit de l'agriculture française de 1700 à 1958. *Cah. ISEA (Serie AF)* **2**, 1–221.

Toutain, J. C. 1967. Les transports en France de 1830 à 1965. *Cah. ISEA (Economies et Sociétés)* **1**(8), 1–306.

Toutain, J. C. 1971. La consommation alimentaire en France de 1789 à 1964. *Cah. ISEA (Economies et Sociétés)* **5**(2), 1909–2049.

Tracy, M. 1964. *Agriculture in western Europe; crisis and adaptation since 1880.* London: Cape.

Tresse, B. T. 1803. Aperçu général du département de Seine-et-Marne. *AS* **6**, 85–90.

Tresse, R. 1955. Le développement de la fabrication des faux en France de 1785 à 1827 et ses conséquences sur la pratique des moissons. *AESC* **10**, 341–58.

Tribondeau, J. 1937. *L'agriculture du Pas-de-Calais.* Arras: Malfait.

Trouvé, M. 1819. Description générale et statistique du département de l'Aude. *AAF* **6**, 384–408.

Viallon, J. B. 1977. L'agriculture bourguignonne. *AB* **49**, 135–42.

Vidal de la Blache, P. 1903. *Tableau de la géographie de la France.* Paris: Hachette.

Vidalenc, J. 1952. L'approvisionnement de Paris en viande sous l'ancien régime. *RHES* **30**, 116–32.

Vidalenc, J. 1965. Les habitants de l'Orne au début de la monarchie de Juillet. *AN* **15**, 567–85.

Vidalenc, J. 1970. *Le peuple des campagnes; la société française de 1815 à 1848.* Paris: Rivière.

Vigier, P. 1963. *La Seconde République dans la région alpine.* Paris: Presses Universitaires de France.

Villeneuve, H. and E. Robert 1839. Revue agricole de la Provence. *Annales Provençales d'Agriculture Pratique et d'Economie Rurale* **12**, 25–93.

Vilmorin, H. 1880. *Les meilleurs blés, description et culture des principales variétés de froments d'hiver et de printemps.* Paris.

Walton, J. R. 1979. *Mechanisation in agriculture: a study of the adoption process.* Inst. Br. Geog. Spec. Publn **10**, 23–42.

Weber, E. 1977. *Peasants into Frenchmen; the modernisation of rural France 1870–1914.* London: Chatto and Windus.

Winchester, H. P. M. 1980. *Rural and urban perspectives on population mobility in France.* Unpubl. D.Phil. thesis, University of Oxford.

Wolff, P. (ed.) 1967. *Histoire du Languedoc.* Toulouse: Privat.

Wolkowitsch, M. 1960. *L'économie régionale des transports dans le centre et le centre-ouest de la France*. Paris: SEDES.

Wright, G. 1964. *Rural revolution in France; the peasantry in the twentieth century*. Stanford: Stanford University Press.

Yvart, J. A. V. 1819. *Excursion agronomique en Auvergne, principalement aux environs des Monts d'Or et du Puy-de-Dôme*. Paris.

Zeldin, T. 1973, 1977. *France, 1848–1945*, 2 vols. Oxford: Clarendon Press.

Zolla, D. 1885. Etude sur la diminution du nombre des ovides en France et en Europe. *AA* **11**, 253–71.

Zolla, D. 1887. Du rôle de la grande propriété et de l'étendue des exploitations agricoles en France et en Angleterre. *AA* **13**, 145–66.

Zolla, D. 1888. Etude sur l'enquête agricole de 1882. *AA* **14**, 433–65.

Zolla, D. 1899. La baisse des prix et la crise agricole. *AA* **25**, 49–75, 145–77, 356–87, 459–91.

Zolla, D. 1920. *L'agriculture moderne*. Paris: Flammarion.

# Index